C H I N A 2

CLEAR WATER, BLUE SKIES

microinfo ltd
P.O. Box 3
Omega Park
Alton
Hampshire
GU34 2PG
United Kingdom

Tel: 01420 86848   Fax: 01420 89889
e-mail: ops@ukminfo.demon.co.uk
URL: http://www.microinfo.co.uk/

CLEAR WATER, BLUE SKIES

CHINA'S ENVIRONMENT IN THE NEW CENTURY

THE WORLD BANK
WASHINGTON D.C.

Contents

This report uses *Hong Kong* when referring to the Hong Kong
Special Administrative Region, People's Republic of China.

Acknowledgments

his report was written by Todd M. Johnson, Feng Liu, and Richard Newfarmer under the direction of Richard Newfarmer (task manager). The report benefited from the contributions of a World Bank team that included Ramesh Ramankutty and Jitu Shah (acid rain); Biliang Hu (economist); Susmita Dasgupta, Hua Wang, and David Wheeler (industry); Gordon Hughes and Kseniya Lvovsky (macro-environmental projections); Robin Carruthers and Tilly Chang (transportation); Leila Cruz and Jennifer Solotaroff (task assistants); and Natasha Berschorner, Daniel Gunaratnam, and Chantale Wong (water).

The report also includes findings from papers prepared by Jonathan Sinton (indoor air pollution); David Fridley (lead); Zhongxiao Jiang, Laurence Lau, and Chen Xikang (macroeconomic projections); Helen Chan (ODS); and Noureddine Berrah, Masaki Takahashi, and Stratos Tavourlareas (power).

In addition, the report benefited from the close collaboration of China's National Environmental Protection Agency and State Planning Commission. The authors are particularly grateful to Zhang Kunmin, Deputy Administrator of the National Environmental Protection Agency, for his thoughtful comments and insights, as well to the Foreign Economic Cooperation Office of the National Environmental Protection Agency in coordinating the counterpart team from the Chinese government. The authors also thank the representatives of various ministries and institutions for their detailed comments on drafts of this report. The Canadian International Development Association facilitated this collaboration by providing resources for a workshop held in Beijing midway through the preparation of this report.

The authors are also indebted to a number of Chinese scholars who wrote background papers for this report: Cao Fengzhong, Guo Xiaomin, Hao Jiming, Liu Qifeng, Wang Hanchen, Yang Zhifeng, Zheng Yisheng, and Zhou Jin.

Several reviewers offered valuable comments at different stages during the preparation of the report, including Richard Ackermann, Theodore Atwood, John Briscoe, William Chandler, Maureen Cropper, Devra Davis, David Fridley, Joseph Goldberg, Nicholas C. Hope, Kathie Krumm, Jane Loos, Peter Midgley, Vikram Nehru, Karen Polenske, Zmarak Shalizi, Jonathan Sinton, Andrew Steer, Robert Taylor, Lee Travers, and Robert Watson.

Walter Spofford of Resources for the Future made a special contribution before he passed away in November 1996. Not only was he a peer reviewer for the concept paper of this report, but his prior work on control of industrial pollution provided much of the rich detail of our knowledge in this important area. Finally, Nicholas C. Hope, former Director of the China and Mongolia Department, provided the inspiration and impetus for the China 2020 series. The authors are grateful for his invaluable comments and insights into the economic issues facing China, as well as for his unwavering support for this activity.

The report was edited by Paul Holtz, laid out by Glenn McGrath, and designed by Kim Bieler, all with the American Writing Division of Communications Development Incorporated.

Overview

Robust economic growth has brought immeasurable benefits to the Chinese people. Incomes have increased, poverty has fallen, and health indicators have improved. Yet the same unbridled growth that has lifted millions out of poverty has also caused serious environmental damage.

Over the past decade China has begun to curb this damage. New laws have established comprehensive regulations for the environment. The government has invested considerable resources in protecting air and water. And economic reforms have reduced the economy's energy intensity at an impressive pace. Although ambient concentrations of most pollutants are still unacceptably high, they have been kept in check.

Chinese leaders know that much remains to be done. Particulate and sulfur levels in major Chinese cities are among the highest in the world, exceeding World Health Organization and Chinese standards by two to five times.

Pollution is one reason chronic obstructive pulmonary disease—emphysema and chronic bronchitis—has become the leading cause of death in China, with a mortality rate five times that in the United States. As Premier Li Peng noted last year, China not only has "to provide people with riches and material products, but also gradually improve the quality of life, and environment is an important part of the quality of life."

This report's findings underscore the environmental concerns of Chinese leaders:

• An estimated 178,000 people in major cities suffer premature deaths each year because of pollution.

• Indoor air pollution, primarily from burning coal and biomass for cooking and heating, causes 111,000 premature deaths each year, mainly in rural areas.

• Each year some 7.4 million work-years are lost to health damages related to air pollution.

• Water pollution, a focus of recent Chinese policy, has contaminated 52 of 135 monitored urban river sections. These sections do not even meet the lowest standards necessary for irrigation water, rendering them mere waste sinks. Unless action is taken, future access to safe drinking water is threatened for tens of millions of people.

• Acid rain in the high-sulfur coal regions of southern and southwestern China threatens to damage 10 percent of the land area, and may already have reduced crop and forestry productivity by 3 percent.

• Children in Shenyang, Shanghai, and other major cities have blood-lead levels averaging 80 percent higher than levels considered dangerous to mental development.

• Air and water pollution damages, especially the dangers fine airborne particulates pose to human health, have been estimated to be at least $54 billion a year— or nearly 8 percent of GDP in 1995.[1]

Despite the magnitude of the problem, China has an unprecedented opportunity to increase its environmental quality of life. Rapid economic growth makes clearer waters and bluer skies more attainable. High rates of investment can be used to develop cleaner, more energy-efficient industries. Policies that channel investment into cleaner production, encourage material and energy efficiency, and encourage conservation of scarce resources could reduce emissions in 2020 below today's levels, improve air and water quality, and lower pollution-related health costs by 75 percent—even as China quadruples its output.

But these outcomes will not happen automatically. Indeed, they will require considerable effort. Although structural change and new technology will help reduce pollution per unit of output, without new policies the pressures from a growing economy will swamp these improvements, and may even undermine recent gains. The old pattern of growth, projected into the future, would generate rising health costs and dramatically raise the cost of cleanup to future generations.

A new growth strategy should aspire to:

• Increase substitution of cleaner fuels—especially natural gas—for coal in household cooking and heating.

• Improve energy efficiency and diversify energy supplies into noncoal sources.

• Reduce emissions from industrial boilers and furnaces.

• Curb indoor air pollution in rural households.

• Increase wastewater collection and treatment from cities and towns and industrial enterprises, especially small enterprises.

• Control increasing acid rain deposition in sensitive regions.

• Prevent automobile-related air pollution.

Since current policies are fragmented across several government agencies with differing policy mandates, a serious assault on China's pollution problems will require not only more aggressive policies, but also better integration of policies and implementation efforts.

China's increasing market orientation requires a strategy for future environmental protection that goes beyond the command and control measures of the past. The next generation of policy should be based on three principles:

• *China must harness the market to work for the environment, not against it.* This means accelerating economic reforms—reorienting state enterprises so that they respond to environmental penalties, adjusting the pricing system to ensure that it reflects environmental costs, liberalizing international trade to give Chinese industry access to the latest environmental technology, and developing capital markets to provide financing to firms and municipalities supplying environmental infrastructure. Harnessing the market also means recognizing that China, with its limited resource base, is underpricing energy and water. Prices of natural gas and water, for example, should be raised to reflect their scarcity. Finally, harnessing the market means that the government should expand the use of taxes on pollution to incorporate its enormous social costs. The real cost to society of coal—after its damages to human health are included—is 100 percent higher in Beijing than its current price. Environmental taxes on coal and gasoline and on pollutants would use the market to clean the environment.

• *China must harness growth for the environment by pursuing investments with the highest environmental benefits for future generations.* Better pricing makes investments in clean technology economic and imposes environmental discipline on firms. For example, Chinese industry today operates at 50–85 percent of the energy efficiency of OECD countries. Pricing energy correctly would create incentives for firms—state and nonstate—to invest in more efficient technology and abatement. If emission taxes and coal transport charges were correct, investment in coal washing—a priority to curb particulate emissions—would be more economic.

Beyond this, public investments in natural gas to replace home use of coal would increase energy efficiency by two to five times and reduce urban air pollution. Public investments in research and development are critical for bringing new industrial technologies (such as efficient industrial boilers and renewable energy) to the market. Public investment in wastewater treatment is unavoidable, and would be more financially viable if sewerage charges were universally introduced and gradually raised to full cost recovery levels. Investments in public transit systems would help avoid the emergence of automobile-based urban transportation systems, which lead to congestion and heavy pollution. Finally, public investments in wastewater systems, financed through increased cost recovery and better pricing, are crucial for conserving water resources and for cleaning China's rivers, lakes, and coastal waters.

• *China must harness its administrative capabilities for the environment.* At the national level, better regulations and policy coordination could replace direct investment controls. Some regulations, such as phasing out lead from gasoline by 2000, are essential for environmental improvement. As markets grow, regulations should spread to cover township and village industrial enterprises, which are an increasingly large source of pollution but are effectively exempt from regulation. If the government uses its powers to set national standards—for automobile and motorcycle emissions, the energy efficiency of buildings, and so on—it will achieve major improvements in environmental quality.

At the regional level, effective management of water basins and airsheds requires setting up regional authorities to overcome the paralysis afflicting current policymaking. Similarly, regulation of sulfur emissions, like that in the Air Pollution Prevention and Control Law of 1995, must target areas suffering from acid rain to have the maximum effect and to minimize control costs.

At the urban level, environmental master plans could help remove dirty industries from downtown areas and allow for effective urban transportation, two measures that will reduce health damages and pollution loads in the years to come. By 2020 well-organized cities with good public transit could have half as many cars clogging their roads as poorly managed cities without public transit. Finally, environmental education, already under way, will increase citizen demands for a cleaner environment. These demands must find expression in wider community participation in environmental policymaking.

Achieving these goals will require some sacrifices in the near term. Pollution control investments will have to double, to about 1 percent of GDP. Still, these costs are small relative to China's annual investments, which total 35 percent of GDP. They are also small relative to the future costs of cleanup if action is delayed. Most important, today's sacrifices are small relative to the enormous improvements in the quality of life that will accrue to future generations.

If China manages to grow by 6–7 percent a year for the next two decades or so, by 2020 its population will enjoy incomes roughly equivalent to those in Portugal today. New policies and careful investments made today mean China's children and grandchildren would also enjoy clear water and blue skies.

Note

1. Chinese researchers prefer to use the "human capital" approach to mortality and health risk valuation, which produces a damage estimate of about 3.5 percent of GDP (see chapter 2).

chapter one

Growth and China's Environment

Since 1978 China's economic reforms have progressively unleashed market forces and dramatically improved the material well-being of most Chinese citizens. The transition from a command to a market economy has produced economic growth rates that put China among the world's five fastest growing economies. Over the past eighteen years rising incomes have eased poverty, reduced infant mortality, improved child and maternal health, and lengthened life expectancy.

During the same period, however, China's environment has deteriorated significantly. Rapid urbanization and industrialization have generated enormous volumes of air and water pollutants, lowering air and water quality. Between 1980 and 1995 China's urban population soared from 191 million to 352 million people, moving more Chinese closer to industrial smokestacks and residential emissions and increasing the number of people exposed to polluted urban air and water.

Today most Chinese cities are covered by blankets of harmful airborne particulates. In major cities levels of total suspended particulates and sulfur dioxide are two to five times World Health Organization (WHO) guidelines (figure 1.1). Since 1980 surface water and groundwater have grown increasingly polluted because of increased emissions of industrial waste, municipal sewage, and agricultural runoff. Urbanization has tightened already short water supplies, particularly in northern China, and has strained urban sewage treatment capacity. In 1996 less than 7 percent of municipal wastewater was treated. Ambient water quality, especially in urban areas, has suffered. And the government recently reported that acid rain affects 29 percent of China's land area (Hao and Liu 1997).

Over the years high levels of pollution have prompted the government to take corrective actions and expand environmental policies (box 1.1). Still, much remains to be done. This report focuses on two of China's most pressing environmental concerns—air and water pollution—and their relation to economic growth. Specifically, it addresses three questions: What are the costs of pollution in China today? Will future economic growth impair or improve air and water quality? And what policies are needed now to ensure that rising incomes translate into a higher environmental standard of living for current and future generations?

Limited resources, better policies

With 22 percent of the world's population living on just 7 percent of the world's arable land, China's per capita resource base is unusually low. Agricultural land per capita is 28 percent of the world average. Range lands per capita are less than half the world average. And forests and wilderness areas per capita are only 15 percent of the world average (World Bank 1992, p. 98). Similarly, water resources per capita are about one-third the world average, and energy resources (particularly hydrocarbons) are low as well.

The Chinese government knows that these limited resources could eventually constrain economic growth and improvements in welfare. It has recognized that China must shift from an extensive and resource-intensive development pattern to an efficient and sustainable one. The government has implemented a comprehen-

sive legal and administrative framework to protect the environment (box 1.2). Moreover, increasing resources are being invested in environmental protection, with record amounts planned during the Ninth Five-Year Plan. Finally, and perhaps most important, economic reforms have helped increase the allocative efficiency of the economy.

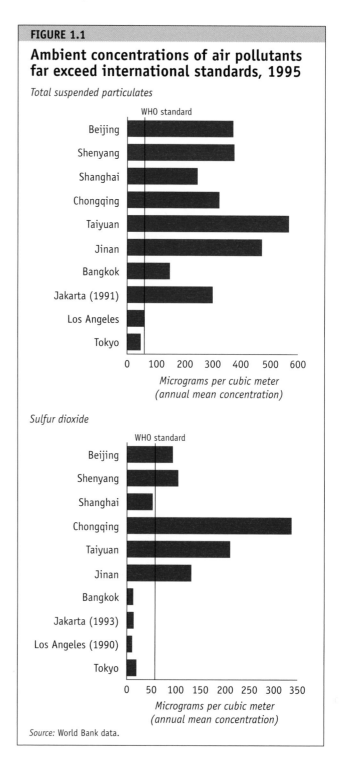

FIGURE 1.1

Ambient concentrations of air pollutants far exceed international standards, 1995

Total suspended particulates

Sulfur dioxide

Source: World Bank data.

Government efforts to address China's environmental problems, 1979–97

1979 Environmental Protection Law (Trial) establishes a general legal framework for regulating environmental degradation

1982 Marine Environment Protection Law promulgated

1983 Government declares environmental protection a basic national policy and formulates three policy principles for controlling pollution

1984 Water Pollution Prevention and Control Law promulgated

1987 Air Pollution Prevention and Control Law promulgated

1989 Environmental Protection Law revised and formally promulgated, and eight environmental management programs endorsed

1992 China among the first nations to act on the Rio agenda: issues "Ten Strategic Policies for Development and the Environment"

China Environmental Protection Action Plan formulated

1994 State Council approves "China Agenda 21—White Paper on Population, Environment and Development," which lays out major policies for sustainable development

1995 Amended Air Pollution Prevention and Control Law includes tougher regulations for controlling sulfur dioxide

Solid Waste Pollution Prevention and Control Law promulgated

1996 State Council approves "Ninth Five-Year Plan for Environmental Protection for 2010," a first in national planning history, and two supplement documents— "Total Emission Quantity Control Plan for Major Pollutants" and "Trans-century Green-engineering Plan"

Amended Water Pollution Prevention and Control Law adds provisions for controlling river basin pollution and imposes tougher regulations for protecting sources of drinking water

Fourth National Environmental Protection Conference: Party Secretary Jiang Zemin and Premier Li Peng deliver keynote speeches calling for more stringent pollution control

State Council issues Decisions Regarding Several Issues of Environmental Protection, laying out guidelines for achieving the environmental targets of the Ninth Five-Year Plan

Government initiates a nationwide campaign to close heavily polluting township and village enterprises; to date more than 60,000 have been closed

1997 Government announces that China will phase out leaded gasoline by 2000; implementation has begun in Beijing

Source: World Bank staff.

As identified by China's National Environmental Protection Agency (NEPA 1993), several environmental priorities are likely to have long-term impacts on ecosystems and the economy:

• Water pollution
• Water shortages in urban and northern China
• Urban air pollution
• Industrial toxic and hazardous wastes
• Soil erosion
• Forest and grassland degradation
• Habitat destruction and species loss.

In a speech to the Fourth National Environmental Conference in 1996, Premier Li Peng noted the environmental achievements that China had made, but admonished the nation that much remained to be done:

We clearly are aware that the situation of the environment in our country is still quite severe. Since China is now still at the stage of rapid industrialization and urbanization, the intensity of exploitation of natural resources rises continuously. In addition, because of the extensive mode of economic growth and backward technological and managerial levels, emissions of pollutants increase continuously. Environmental pollution in center cities is worsening and is extending into rural areas, and the scope of ecological damage is increasing. (NEPA 1996b, p. 9)

Because air and water pollution are the most damaging to human health, worker productivity, and agricultural output, they are the focus of this report.[1] The rest of this chapter identifies the sources of pollutant emissions and their effects on ambient air and water quality.

Air pollution

Many cities in China, especially large ones, have seen ambient air quality improve thanks to industrial emission control and residential fuel switching programs. But ambient concentrations of suspended particulates are still extremely high in most cities, and ambient sul-

BOX 1.2

Environmental regulation in China

Since environmental regulations were first enacted in 1979, China has continued to improve its regulatory practices and to learn from the experiences of other countries. Current environmental legislation covers the management of wastewater, solid waste, and atmospheric and noise pollution; the conservation of oceans and watercourses; and the management and protection of forests, grasslands, soil, fisheries, mineral resources, water resources, wildlife, and coal and other energy resources.

At the national level, environmental policies are formulated by the State Environmental Protection Commission and approved by the State Council. The National Environmental Protection Agency, established in 1988 as an independent agency, is the secretariat for the State Environmental Protection Commission. The agency's primary functions are to disseminate national environmental policy and regulations, collect environmental data from and provide training and administrative support to local environmental protection bureaus, and provide technical and policy advice to the State Environmental Protection Commission on both national and international environmental issues. Implementation of environmental policy in China is also the responsibility of ministries and related agencies at all levels of government.

As in many countries, regulatory enforcement has been the weak link in China's system. In recent years the National Environmental Protection Agency and national ministries have been closely monitoring the enforcement of environmental laws. But to be successful, monitoring and enforcement must take place at the local level. Local governments face a tradeoff between protecting the environment and safeguarding the financial and employment performance of local firms. As a result pollution generally exceeds China's own standards. But local governments share with firms an interest in finding financially attractive pollution abatement investments—in most cases focusing on process changes rather than end-of-pipe controls.

Citizens affected by pollution damage can appeal directly to polluting factories for compensation. Failing a satisfactory resolution of the conflict, the matter is brought to the local environmental protection bureau. Although large pollution accidents and intractable cases find their way to local or higher-level bureaus, thousands of small environmental cases are negotiated directly because it is more efficient and because of the traditional Chinese disdain for legal or administrative arbitration. By encouraging the public to be more vigilant about the environment and to report violations promptly, the enforcement record of local environmental agencies has improved.

Source: World Bank staff.

fur dioxide concentrations and acid rain are also high in areas where high-sulfur coal is consumed. In general, northern cities have more serious particulate pollution because coal is used for space heating, while southern cities have serious sulfur dioxide pollution.

Sources of air pollutants

Rapid growth and rising incomes have increased China's demand for energy. As a result coal use has more than doubled since 1980, reaching some 1.3 billion tons in 1995. Energy consumption—coal, biomass, petroleum products—is responsible for most particulate emissions. Coal combustion in industrial boilers and small household stoves accounts for a disproportionately large percentage of ambient levels of fine (less than 10 microns) and ultrafine (less than 2.5 microns) particles, which are the most damaging to human health. Industrial boilers and household stoves are also responsible for most sulfur dioxide and nitrogen oxide emissions. Industrial boilers, many run inefficiently and with low smokestacks, consume 33 percent of China's

coal and account for an even larger percentage of ground-level air pollution. Although households account for just 15 percent of coal use, they typically contribute 30 percent of the ground-level air pollution in Chinese cities. The average sulfur content of coal produced and consumed in the southwestern provinces of Sichuan, Guizhou, and Guangxi is about 4 percent.

Over the past decade the number of trucks and automobiles in major cities has increased dramatically, creating a new source—though still relatively limited in most of the country—of urban air pollution. Motor vehicles emit particulates, sulfur, carbon monoxide, nitrogen oxides, and volatile organic compounds.

Emissions and concentrations

Particulate emissions have remained relatively flat since the early 1980s, implying a substantial increase in controls given the near doubling of coal consumption (figure 1.2). In contrast, sulfur dioxide emissions have roughly paralleled the increase in coal consumption because of inadequate sulfur control measures.

FIGURE 1.2

Emissions of major pollutants have grown much slower than GDP, 1985–95

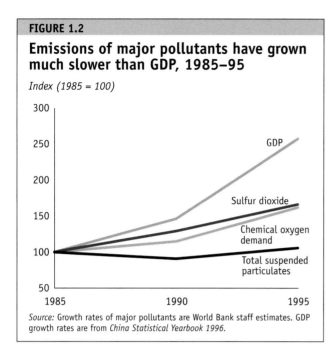

Index (1985 = 100)

Source: Growth rates of major pollutants are World Bank staff estimates. GDP growth rates are from *China Statistical Yearbook 1996*.

dential gas use, expanded district heating, taller stacks on industrial boilers, and urban planning policies that have moved some industries out of central city districts.

Although ambient particulate concentrations in large cities fell significantly between 1985 and 1991, medium-size and small cities experienced less dramatic declines, and have even seen a slight increase since the early 1990s (figure 1.3). Despite the steady increase in sulfur dioxide emissions, average ambient sulfur dioxide concentrations have remained stable in medium-size and small cities and declined in large cities (figure 1.4), presumably due to residential fuel switching and taller stacks. On average, populations in large cities are exposed to 30 percent higher particulate concentrations and 60 percent higher sulfur dioxide levels than populations in medium-size and small cities.

Indoor air pollution is a severe problem in China. A large portion of the population, especially in rural areas, is regularly exposed to pollutant levels well above national ambient standards (table 1.1). Most households still burn solid fuels like raw coal and wood for cooking and heating. Indoor combustion produces particulates, sulfur and nitrogen oxides, carbon monoxide, and other pollutants. These indoor air pollutants contribute to high rates of respiratory disease, the leading cause of death in rural areas and the third leading cause in cities.

Ambient concentrations of particulates and sulfur dioxide in many Chinese cities are among the highest in the world and are significantly above WHO guidelines and Chinese air quality standards. Still, average ambient particulate concentrations in monitored cities have decreased markedly since the mid-1980s, reflecting increased particulate control. In Heilongjiang Province falling ambient concentrations of particulates in a number of cities are reportedly due to increased resi-

FIGURE 1.3

Average ambient particulate concentrations have fallen, 1985–95

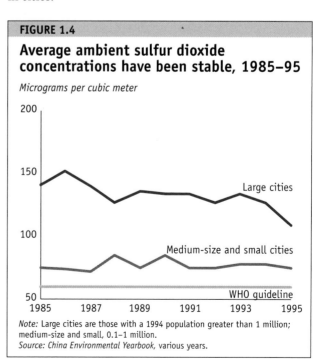

Micrograms per cubic meter

Note: Large cities are those with a 1994 population greater than 1 million; medium-size and small, 0.1–1 million.
Source: China Environmental Yearbook, various years.

FIGURE 1.4

Average ambient sulfur dioxide concentrations have been stable, 1985–95

Micrograms per cubic meter

Note: Large cities are those with a 1994 population greater than 1 million; medium-size and small, 0.1–1 million.
Source: China Environmental Yearbook, various years.

TABLE 1.1

Chinese ambient air quality standards are comparable to those of the World Health Organization
(micrograms per cubic meter)

Pollutant	Averaging time	China Class 1	China Class 2	China Class 3	World Health Organization
Sulfur dioxide	Annual	20	60	100	40–60
Total suspended particulates	Daily	150	300	500	150–230[b]
	Annual[a]	60	120	150	60–90
PM-10	Daily	75	150	250	70
	Annual[a]	20	60	100	
Carbon monoxide	Daily	4	4	6	10[c]
Nitrogen oxides	Daily	50	100	150	150
Ozone	8 hours				100-120
Lead	Annual	0.7			0.5–1.0

Note: Class 1 are tourist, historic, and conservation areas. Class 2 are residential urban and rural areas. Class 3 are industrial areas and heavy traffic areas.
a. Since China does not have annual standards for total suspended particulates, it has been assumed that the same ratio between daily and annual standards for sulfur dioxide applies to total suspended particulates. Annual standards are needed for comparability with ambient concentration data.
b. Guideline values for combined exposure to sulfur dioxide and total suspended particulates.
c. 8 hours.
Source: WHO and UNEP 1992; World Bank staff estimates.

Indoor air quality in many urban households has improved significantly as gas, electricity, and central heating have replaced solid fuels and coal stoves and as kitchen ventilation has improved. About one-third of the urban population now has access to gas for cooking. Households that continue to burn coal increasingly use cleaner and more efficient briquettes in specially designed stoves. Use of briquettes and liquefied petroleum gas (LPG) is spreading to many suburban areas as well. Even biomass is used more cleanly, thanks to a Ministry of Agriculture program that provides efficient biomass stoves with flues to rural households.

Lead is an especially pernicious pollutant because of its irreversible effects on children's intelligence and aptitudes. Although detailed statistics on lead sources are lacking, the main sources of lead in air, water, and soil are probably industrial emissions from mining, lead smelting, production of lead-acid batteries and cables, and coal use. However, leaded gasoline is a growing source of lead emissions. China's gasoline-powered vehicle fleet—cars, motorcycles, and a large percentage of trucks—has grown by more than 10 percent a year since the mid-1980s, and gasoline consumption jumped from about 14 million tons in 1985 to 27 million tons in 1994. About half of China's gasoline production is leaded, primarily the higher-octane fuel used in high-compression engine automobiles.

Two studies from Fuzhou, Fujian Province, have demonstrated a correlation between automobile emissions and elevated blood-lead levels in children (Li, Gu,

and Wang 1992; Li, Gu, and Yu 1994). Other urban areas in China also have reported increases in blood-lead levels as automobile use has increased. Studies conducted in Shanghai show that a newborn's risk of lead poisoning is associated with a number of factors, including a family member's exposure to lead at work, proximity to a major traffic way, and household and neighborhood coal combustion (Shen and others 1996). Nontransportation sources of lead are also important. For instance, ambient lead concentrations in Tianjin are nearly twice as high during the heating season as during warmer months.

Anecdotal evidence suggests that ambient lead levels are below Chinese and WHO standards in most Chinese cities but that levels have been rising over the past decade as vehicle use has increased. In Guangzhou the average ambient lead concentration increased twelvefold between 1988 and 1995, reaching 0.113 micrograms per cubic meter ($\mu g/m^3$). In a heavy traffic area of Taiyuan, Shanxi Province, ambient lead concentrations nearly doubled in three years, from 0.625 $\mu g/m^3$ in 1990 to 1.803 $\mu g/m^3$ in 1993 (Chen 1995).

In addition to local effects, air pollution has regional and global consequences. Long-range transport of acid rain, primarily from sulfur, has already caused serious damage to agriculture, ecosystems, and materials (Hao and Liu 1997). In response, the National Environmental Protection Agency recently recommended establishing sulfur dioxide and acid rain control areas.

Global effects of air pollution include changes in climate and destruction of the ozone layer. China became the world's largest producer of ozone-depleting substances in 1995, after the production of such substances was banned in industrial countries. By the early twenty-first century China is expected to be one of the largest emitters of greenhouse gases (Johnson and others 1996). China attaches great importance to these and other global environmental concerns and was an early signatory of the Climate Change Convention and the Montreal Protocol (box 1.3).

Water pollution

Since 1980 the quality of China's surface water and groundwater has deteriorated significantly under the pressure of rapid industrial development, brisk population and urban growth, and increased use of chemical fertilizers and pesticides. As a result water pollution is now a serious problem for urban and rural drinking water. About 40 percent of the monitored river sections flowing past cities do not meet minimum water quality standards (NEPA 1996a).

Water shortages compound China's urban water quality problems. Low water prices, especially for agriculture, aggravate shortages and increase pollution. The lack of coordinated water basin management also has contributed to surface water pollution in many localities. For example, during dry periods local water use agencies often close gates along their section of a river to retain water for local users, but in so doing retain pollutants from industrial, municipal, and irrigation discharges. When the first rains come, gate operators flush their sections of the river by forcing a highly polluted waste stream into the main river channel. A surge of this sort on the Huai River in 1994 killed massive numbers of fish, caused widespread illness, and forced municipal and industrial water intakes along the river to shut down.

Sources

Nationwide, municipal waste accounts for about 40 percent of wastewater discharges; industry contributes the rest (figure 1.5). In the Huai River Basin, which runs through Anhui and Jiangsu provinces, industrial discharges account for about 75 percent, and municipalities the remaining 25 percent, of point-source chemical oxygen demand pollution. Despite limited data, small-scale activities—township and village enterprises and

---

BOX 1.3

Protecting the ozone layer—a model of international cooperation

Ozone-depleting substances comprise chlorofluorocarbons and halons, which attack and deplete the stratospheric ozone layer. Without the protection of the ozone layer, ultraviolet radiation reaches the earth's surface, increasing the incidence of skin cancers and eye disease, contributing to global warming, reducing agricultural yields, destroying natural vegetation, and damaging aquatic organisms. Ozone-depleting substances are used in many common consumer products, including aerosols, coolants, foams, fire extinguishers, insulation, automobile air conditioners, refrigerators, and solvents.

With rapid industrialization and urbanization, China's consumption of ozone-depleting substances grew by more than 12 percent a year between 1986 and 1994. By the end of 1994 China produced about 60,000 tons and consumed about 81,000 tons of these substances.

The Chinese government ratified the Montreal Protocol on Substances that Deplete the Ozone Layer in June 1991, committing to phase out production and consumption of ozone-depleting substances by 2010. These efforts are eligible for financing from the Multilateral Fund, created to help developing countries meet their Montreal Protocol obligations. China committed to reducing consumption of ozone-depleting substances in 1996 to the 1991 level of 50,000 tons; it will then reduce that level by half by 2000 and phase out consumption completely by 2010. With rapid economic growth and a slow start in phaseout, the 1996 target was not met. But with the necessary regulations now in place, the government is confident that it can meet its future obligations.

As the implementing agency for the Montreal Protocol in China, the World Bank has been working on phaseout activities in China since 1991. When its first three Montreal Protocol projects are completed, about 25 percent of China's potential ozone-depleting production and consumption will have been eliminated. A fourth project will target halons, which account for 20 percent of China's consumption of ozone-depleting substances. Most halons will then be eliminated within three years. China also plans to ban the use of chlorofluorocarbons in aerosols in 1998.

Source: World Bank staff.

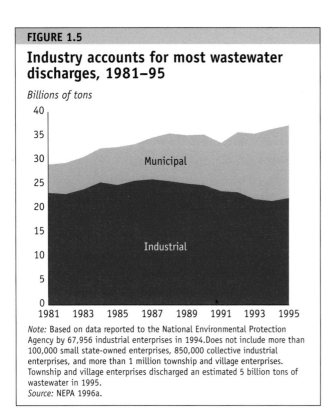

FIGURE 1.5

Industry accounts for most wastewater discharges, 1981–95

Billions of tons

Note: Based on data reported to the National Environmental Protection Agency by 67,956 industrial enterprises in 1994. Does not include more than 100,000 small state-owned enterprises, 850,000 collective industrial enterprises, and more than 1 million township and village enterprises. Township and village enterprises discharged an estimated 5 billion tons of wastewater in 1995.
Source: NEPA 1996a.

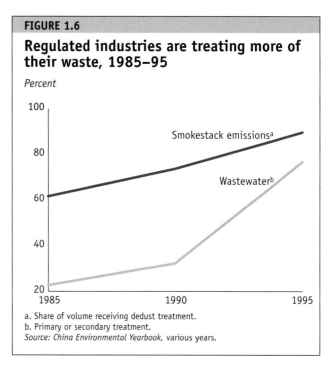

FIGURE 1.6

Regulated industries are treating more of their waste, 1985–95

Percent

a. Share of volume receiving dedust treatment.
b. Primary or secondary treatment.
Source: China Environmental Yearbook, various years.

small industries, coal mines, and agriculture—are known to be significant contributors to water pollution problems in adjacent surface waters and groundwaters.[2]

Industrial waste. Pulp and paper, metallurgical, chemical, and mining plants account for most industrial water pollution.[3] In addition, agroprocessing and food industries—grain processing, breweries, leather, and tanning—are often a major source of chemical oxygen demand and biological oxygen demand pollution.

Municipal waste. Although 77 percent of industrial wastewater from regulated industries receives some treatment (figure 1.6), only 7 percent of municipal wastewater is treated. And while industrial wastewater discharges have been decreasing over the past decade, municipal wastewater discharges have been increasing. The contribution of municipal wastewater to chemical oxygen demand and biological oxygen demand pollution varies by region. For example, municipal waste accounts for only 40 percent of wastewater discharges in Shandong, but accounts for 68 percent of chemical oxygen demand pollution and 85 percent of biological oxygen demand pollution in Chongqing.

Agriculture. In addition to the indirect water pollution generated by agroprocessing industries, agriculture is responsible for significant water pollution from crop production and animal husbandry. Recent studies suggest that the intensive use of nitrogen fertilizer and pesticides are a serious source of water pollution. Although fertilizer use in China is not high by international standards, its effectiveness is low because of poor-quality fertilizers and the excessive use of nitrogen-based fertilizers relative to phosphorous and potassium fertilizers. Of particular concern is the widespread use of ammonia bicarbonate (ABC) fertilizer, which is cheap and easy to use but is also soluble and easily washed out to streams, lakes, and aquifers. Pesticide use, more widespread in recent years, has been implicated in species loss (birds) and has polluted some important water bodies.[4] Animal waste from livestock farms is another major source of biological oxygen demand and coliform pollution. Meat production (chicken, pork, beef, lamb) has grown considerably over the past fifteen years, and much of the manure from livestock farms has found its way into nearby water systems. A study of Hangzhou Bay concluded that agriculture contributed 88 percent of chemical oxygen demand pollution in 1994, and that nutrients from fertilizer and manure posed the greatest threats to the bay's ecological balance and to inland surface water and groundwater quality (box 1.6).

The poor get more . . . pollution

Since richer areas have more industrial production, they might also be expected to have more pollution. Unfortunately for China's poor, however, even this small consolation is denied them. A comparison of fifty Chinese cities shows that as wages rise, the pollution density of suspended particulates falls (box figure). Poor cities also have higher emissions of sulfur dioxide than rich cities, although emissions density rises from low- to middle-income cities and then falls to the lowest levels in high-income areas.

Why this counterintuitive association between poverty and pollution? Although richer areas do have higher levels of indus-trial production, production in those areas is also much cleaner, for at least three reasons. First, and probably most important, environmental regulation is stricter in rich areas because these communities contain highly educated citizens who are concerned about environmental problems, well informed about them, and organized to defend their interests. Second, production in low-wage communities generates more pollution because industrial facilities with unskilled workers are generally less efficient and produce more waste. Finally, demand in high-income markets tends to favor products whose production generates less pollution.

Low-income cities are more polluted than high-income cities

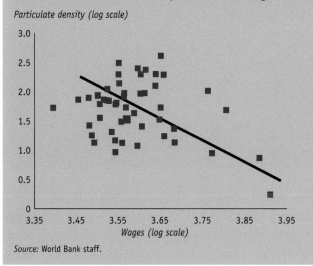

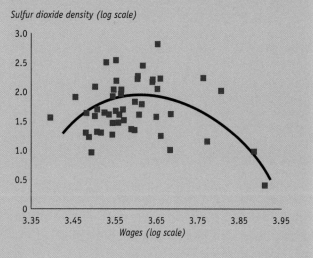

Source: World Bank staff.

Emissions and concentrations

Wastewater discharges from monitored sources increased slightly between 1981 and 1995 (see figure 1.5), but the trend was not consistent for industrial or municipal wastes or for individual provinces or river basins. Industrial wastewater discharges appear to have fallen slightly over this time, though the data do not capture unmonitored township and village enterprises and small industries. Although some provinces (Tianjin, Shanghai, Liaoning, Zhejiang) managed to cut indus-trial discharges leading to chemical oxygen demand pollution, others (Shandong, Guangxi, Xinjiang, Shaanxi) experienced large increases.

Municipal wastewater discharges nearly tripled between 1981 and 1995, increasing by almost 7 percent a year. Discharges from households without sewerage are not included in these data and presumably increased

by similar—if not larger—amounts during the same period.

Surface water quality in urban areas, especially in the water-short northern regions, is poor. About half of the monitored urban river sections in northern China do not meet the lowest ambient standards (grade 5), making the water unsuitable even for irrigation (figure 1.7). Only about 8 percent of the monitored urban river sections in northern China meet the standards for direct human contact (grade 3 and better). Southern urban river sections are less polluted, partly because of the region's larger assimilative capacity. While the Yangtze and Pearl rivers in the south have fairly clean water, the northern Yellow, Huai, Hai and Luan, and Daliao rivers are quite polluted, with large portions of water unsuitable as a raw source of water for potable water systems.

Rivers tend to be polluted with biological oxygen demand, chemical oxygen demand, and petroleum, and water quality near cities and downstream of wastewater discharges is particularly poor. Studies of the Yangtze River have found that the most polluted sections of the upper Yangtze are near Chongqing, where in 1993 indexes for chemical oxygen demand,

chromium, mercury, lead, ammonia nitrogen, petroleum, acidity, and coliform exceeded National Environmental Protection Agency standards. Heavy metal pollution of water is not a widespread problem, though instances of severe metals pollution do exist.[5]

Although water pollution problems are serious, improvements have been made in a number of areas. The water quality of major rivers has remained relatively stable in the 1990s. Industrial wastewater control efforts have reduced wastewater discharges. And large-scale water pollution control campaigns are targeting the most polluted rivers and lakes.

Conclusion

Although the past eighteen years of economic growth have brought many benefits to China, the environment has suffered. Moreover, the large scale and rapid growth of China's economy have exposed a growing population to serious air and water pollution (though in some cases pollution levels have fallen). Without more stringent pollution control, environmental damage will likely worsen. The government has made controlling air

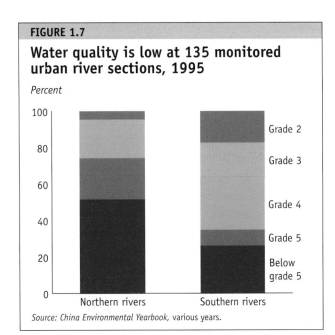

FIGURE 1.7

Water quality is low at 135 monitored urban river sections, 1995

Percent

Source: China Environmental Yearbook, various years.

and water pollution a national priority because pollution damages human health, lowers worker productivity, reduces agricultural output, and disrupts normal ecosystem functions. The value of these losses is the subject of the next chapter.

Notes

1. Environmental degradation costs such as soil erosion, land degradation, deforestation, and desertification are difficult to quantify. Moreover, it is hard to assess natural versus human-made causes. Nonetheless, Chinese and international researchers have estimated some of these costs. Smil (1996), for instance, estimates that annual environmental degradation costs in China amounted to 5 percent of GDP in 1990. The largest of these costs came from the overcutting of mature forests (60 percent of the total) and soil erosion (20 percent).

2. Along the Xiaoqing River Basin, for example, the provincial environmental protection agency monitors only point sources, and relies on other provincial agencies—the construction commission, agricultural bureau, hydraulic bureau—to monitor and estimate discharges from small towns and from agriculture.

3. Of the thirty-six large point-sources that are monitored for water pollution along the Jialing River near Chongqing, six enterprises account for 80 percent of industrial chemical oxygen demand emissions, with two factories (pharmaceuticals and paper) contributing 42 percent of the total; four enterprises (pharmaceuticals, print and dyeing, paper, glue) account for 75 percent of industrial biological oxygen demand emissions; one synthetic chemical plant accounts for all phenol emissions; one specialty steel plant accounts for 57 percent of oil emissions; and one electroplating factory contributes half of all chromium emissions.

4. Although the production and use of DDT has been banned in China since the early 1980s, trace amounts of the pesticide exceeding maximum risk levels were found in the mouth of the Yangtze River in 1995 (Delft Hydraulics, Mott MacDonald, and Welsh Water International 1995). Chinese researchers believe that these DDT concentrations are due to residual amounts of the pesticide remaining in the river sediment.

5. Large waste heaps of metal ore and chemicals at a metallurgy plant along the Yangtze River in Chongqing were found to be contributing serious chromium pollution to the river during rains and periods of high water. Storing dangerous waste along major rivers is common and can pose a significant water pollution hazard. Lead concentrations taken along the Yangtze are higher in low-stream (7.52 μ/l) and mid-stream (4.38 μg/l) than in up-stream (2.76 μg/l) sections.

References

Chen, S.L. 1995. "China Vehicle Pollution Management: Challenges and Prospects." Chongqing Environmental Protection Bureau, Chongqing.

Delft Hydraulics, Mott MacDonald, and Welsh Water International. 1995. "Hangzhou Bay Environmental Study." Report prepared for the World Bank's International Development Association, Washington, D.C.

Hao, J.M., and B.J. Liu. 1997. "Acid Rain in China: Patterns, Trends and Impacts." National Environmental Protection Agency, Beijing.

Johnson, T.M., J.F. Li, Z.X. Jiang, and R.P. Taylor, eds. 1996. *China: Issues and Options in Greenhouse Gas Emissions Control.* World Bank Discussion Paper 330. Washington, D.C.

Li, G.H., W.X. Gu, and D.M. Wang. 1992. "Effect of Lead Pollution on IQ in Children." *China Journal of Preventive Medicine* 26: 380–81.

Li, G.H., W.X. Gu, and M. Yu. 1994. "Effect of Lead Pollution on Blood Pb, Cu, Zn, and SOD in Primary Schoolers." *China Public Health* 10: 204–05.

NEPA (National Environmental Protection Agency). 1993. *Environmental Action Plan of China.* Beijing.

———. 1996a. *China Environmental Yearbook.* Beijing.

———. 1996b. "Selected Documents from the Fourth National Environmental Conference." Beijing.

Shen, X.M., J.F. Rosen, D. Guo, and S.M. Wu. 1996. "Childhood Lead Poisoning in Children." *Science of the Total Environment* 181: 1001–109.

Smil, V. 1996. "Environmental Problems in China: Estimates of Economic Costs." East-West Center Special Report 5. Hawaii.

WHO (World Health Organization) and UNEP (United Nations Environment Programme). 1992. *Urban Air Pollution in Megacities of the World.* Oxford: Blackwell.

World Bank. 1992. *World Development Report 1992: Development and the Environment.* New York: Oxford University Press.

WRI (World Resources Institute). 1995. *World Resources, 1994–95.* New York: Oxford University Press.

chapter two

Pollution, Health, and Productivity: Estimating Today's Costs

ir and water pollution are costly. Polluted air and water can damage the health of people exposed to them, in some cases leading to premature death. Air and water pollution can also lower the productivity of workers, as with chronic bronchitis, diarrhea, or reduced intelligence from childhood lead exposure. Acid rain or water shortages associated with degraded water supplies can reduce agricultural productivity. What are the costs of these damages, and which damages should worry Chinese policymakers the most?

Air pollution

China has a high incidence of respiratory disease. Chronic obstructive pulmonary disease, including emphysema and chronic bronchitis, is the leading cause of death—a disease burden more than twice the average for developing countries (World Bank 1996a).[1] While smoking is the most important proven risk factor for chronic obstructive pulmonary disease in high-income countries, nearly equal rates of the disease among men and women in China (despite the fact that few women smoke) suggest that high concentrations of indoor and outdoor air pollutants are the main cause of chronic obstructive pulmonary disease in China. An exposure-response relationship exists between concentrations of particulates and sulfur dioxide and several health outcomes, including loss of pulmonary function, chronic respiratory illness, bronchitis, premature death, and increased hospital visits and admissions (box 2.1).[2]

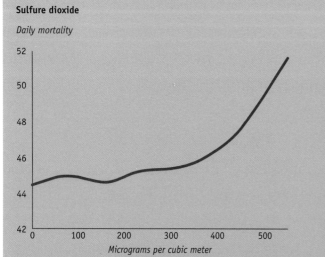
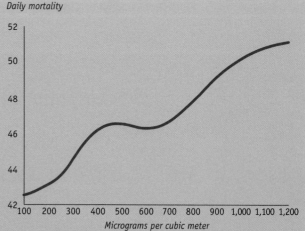

Air pollution in excess of Chinese standards provides a benchmark against which to quantify the effects air pollution has on human health (see table 1.1). To calculate urban air pollution damages, we assumed that average ambient concentrations of particulates and sulfur in Chinese cities are similar to levels in Beijing in 1995 and that 80 percent of China's urban population—240 million people—is exposed to ambient air pollution in the relevant airsheds. Using dose-response functions from studies in China and around the world (see box 2.1), it is estimated that about 178,000 premature deaths could be avoided each year if China met its class 2 air pollution standards (table 2.1).

Urban air pollution's toll on other aspects of human health is also severe. Hospital admissions due to respiratory illness are 346,000 higher because of excess pollution. Moreover, high pollution causes 6.8 million emergency room visits each year. And some 4.5 million person-years are lost because of illnesses associated with pollution levels that exceed standards.

Indoor air is sometimes more polluted than outdoor air. Studies from several parts of China have found that particulate levels (from biomass and coal combustion) are generally higher indoors (above 500 micrograms per cubic meter, or $\mu g/m^3$), and that the highest concentrations (above 1,000 $\mu g/m^3$) are found in kitchens (Sinton and others 1995).

Estimates based on conservative assumptions about indoor air pollution suggest that it causes 111,000 premature deaths a year (see table 2.1). Indoor air pollution also causes deaths from carbon monoxide poisoning in northern China, where coal is used to heat tightly weatherproofed dwellings. The health problems caused by indoor fuel use are on a scale roughly comparable to that posed by smoking (box 2.2). Women and children suffer the highest incidence of disease from indoor air pollution.

As noted in chapter 1, the proliferation of alternative energy sources for cooking and heating—gas, electricity, central heating—is making inroads into this problem. About one-third of urban residents have access to gas for cooking. Moreover, gas supplies are expanding, and more people are moving from single-story brick houses to apartment buildings (which are more likely to have gas). Households that continue to burn coal are increasingly using more efficient, cleaner briquettes in specially designed stoves, and their use is spreading to many suburban areas.

Exposure to lead is particularly damaging because of its effects on intelligence. The most common index of lead exposure is the lead level of blood. While there are no comprehensive national data, studies from various Chinese cities show that blood-lead levels in children in

BOX 2.2

Like people everywhere, the Chinese need to break the smoking habit

Although indoor and ambient air pollution cause enormous damage to human health in China (see table 2.1), cigarette smoking may be even more dangerous. In 1989 smoking generated $5.1 billion in health care costs. In 1990 smoking-related deaths—chronic obstructive pulmonary disease, coronary heart disease, and lung cancer—reached about 1 million, or more than 9 percent of total deaths. Over the next twenty-five years this figure is expected to double. Average annual cigarette consumption in China rose from 730 per person in the early 1970s to 1,290 in the early 1980s, to 1,900 by the 1990s. If current patterns persist, about 50 million Chinese who are now less than 19 years old will die as a result of smoking. For people exposed to high indoor and outdoor air pollution, smoking multiplies the risk of respiratory disease.

Source: World Bank 1996b.

TABLE 2.1

A heavy annual toll in physical damages could be avoided if China met its class 2 air quality standards

Problem	Number of cases averted
Urban air pollution	
Premature deaths	178,000
Respiratory hospital admissions	346,000
Emergency room visits	6,779,000
Lower respiratory infections or child asthma	661,000
Asthma attacks	75,107,000
Chronic bronchitis	1,762,000
Respiratory symptoms	5,270,175,000
Restricted activity days (years)	4,537,000
Indoor air pollution	
Premature deaths	111,000
Respiratory hospital admissions	220,000
Emergency room visits	4,310,000
Lower respiratory infections or child asthma	420,000
Asthma attacks	47,755
Chronic bronchitis	1,121,000
Respiratory symptoms	3,322,631,000
Restricted activity days (years)	2,885,000

Source: World Bank staff estimates.

both polluted and nonpolluted areas significantly exceed the U.S. standard—10 micrograms per deciliter (µg/dl)—above which adverse health effects occur. In Shanghai children living in industrial and congested traffic areas had blood-lead levels averaging 21.8–67.9 µg/dl. Between 65 and 100 percent of children have blood-lead levels above 10 µg/dl (Shen and others 1996).

Studies of children in Shanghai, Shenyang, Fuzhou, and Beijing have consistently found a link between lead exposure and deficits in intelligence, neurobehavioral development, and physical growth. A study of 133 infants in the Yangpu District of Shanghai found a statistically significant difference in mental development (based on standardized tests) between children with high blood-lead levels and children with lower levels. The prospect that millions of Chinese children might be exposed to lead and the productivity of an entire generation of urban workers might be lowered underscores the importance of the government's recent decision about removing lead from gasoline and from the environment.

Water pollution

Water pollution damages human health, fisheries, and agriculture (from polluted irrigation water) and increases spending on clean water supplies. Water pollution also damages ecosystems—through eutrophication and the loss of plant and animal species—though these damages are more difficult to quantify.

Several diseases have been linked to human contact with polluted water, including diarrhea, hepatitis, ascaris, hookworm infection, trachoma, and dracunculiasis. Improvements in water supply and sanitation can substantially reduce the incidence and severity of these diseases, as well as the infant mortality associated with diarrhea.[3] In general, water-related diseases are less common in China than in other developing countries. Relative to other Asian countries at similar incomes levels, China's water supply and sanitation are good, although there are differences between urban and rural areas (table 2.2). In 1990 just 1.5 percent of deaths and 3 percent of the disease burden were the result of common water- and sanitation-related diseases (diarrhea, hepatitis, trachoma, intestinal nematodes). By contrast, chronic obstructive pulmonary disease accounted for 16 percent of deaths and 8.5 percent of the disease burden, while respiratory infections accounted for 5 percent of deaths and nearly 6 percent of the disease burden (table 2.3).

Water pollution is believed to have other effects on health—including skin disease, liver cancer, birth defects, and spontaneous abortion. Although several studies have been carried out in China to assess the relationship between human contact with polluted water and these diseases, it is often difficult to show the exact cause of these often chronic diseases without large epidemiological surveys carried out over many years. In contrast to fecal-borne diseases such as diarrhea, water pollution–related cancers and birth defects are believed to be caused by heavy metals and toxic chemicals that should be removed through industrial primary and pre-treatment of wastewater.

Urban sanitation in China is in transition. Night soil collection for agricultural use has decreased with alternative fertilizer sources and rising rural incomes, the volume of municipal waste has increased with growing urban populations, and modern sewerage systems have not been able to keep up. Thus there is potential for an increase in fecal-borne diseases, particularly in northern China.

TABLE 2.2
China has fairly widespread access to safe drinking water and sanitation, 1990
(percent)

Country	Safe drinking water			Sanitation[a]		
	Urban	Rural	Total	Urban	Rural	Total
China	87	68	72	100	81	85
India	86	69	73	44	3	14
Indonesia	35	33	34	79	30	45
Sri Lanka	80	55	60	68	45	50
Japan	100	85	96	100	100	100

a. Assumes that residents have access to water for washing and that sewage is removed from the house through outdoor latrines, night soil collection systems, or flush toilets.
Source: World Bank 1996b.

TABLE 2.3

Water pollution accounts for a much smaller share of China's mortality and disease burden than air pollution, 1990

Illness	Deaths		Disability-adjusted life-years (DALYs)		China's share of global DALYs[a] (percent)
	Thousands	Share of Chinese total (percent)	Thousands	Share of Chinese total (percent)	
Water pollution–related	134	1.5	6,042	2.9	9.6
Diarrhea	93		3,685		4.3
Hepatitis	34		626		34.7
Trachoma	—		347		14.2
Intestinal nematodes	7		1,384		35.2
Air pollution–related	1,899	21.2	29,734	14.2	17.2
Chronic obstructive pulmonary disease	1,432	16.0	17,810	8.5	58.6
Lower respiratory tract infection	467	5.2	11,924	5.7	9.0
Total	8,885		208,407		15.1

— Not available.

a. In 1994 China's population accounted for 21.3 percent of the world's population.

Source: World Bank 1993 and 1996b.

Regional and global air pollution

Some air pollution problems, such as acid rain and climate change caused by greenhouse gas emissions, are of regional and even global concern.

Acid rain

Acid rain refers to the products of chemical reactions between airborne pollutants (sulfur and nitrogen compounds) and atmospheric water and oxygen. When fossil fuels are burned, oxides of sulfur and nitrogen are emitted into the atmosphere. Once there, sulfur dioxide and nitrogen oxides react with other chemicals to form sulfuric acid and nitric acid. These gaseous emissions can stay in the atmosphere for several days and travel hundreds or thousands of kilometers before falling back to the earth's surface as acid rain. This process is more accurately termed *acid deposition,* since acidity can travel to the earth's surface in many forms: rain, snow, fog, dew, particles (dry deposition), or aerosol gases.

Sulfur emissions are highest in southern China, particularly in Sichuan, Guizhou, Guangxi, and Hunan provinces, primarily because of the large amounts of high-sulfur coal produced and consumed there. The power sector, with its widespread use of tall smokestacks, is probably the main contributor to the long-range transport of sulfur and nitrogen oxides. However,

simulations using the Regional Air Pollution Information and Simulation Model-Asia (RAINS-Asia) model show that a sizable portion of long-distance sulfur deposition comes from nonpower, or "area," sources.[4]

Acid rain is associated with many types of damage, from effects on human health to destruction of boreal and subtropical evergreen forests. Effects on forests and agricultural crops occur directly through high ambient concentrations of sulfur dioxide and nitrogen oxides and indirectly through the acidification of soils. Direct effects occur mainly near the sources of emissions. Soil acidification is likely to have long-term impacts on ecosystems and is believed to be the cause of the extensive dieback of forests in Central Europe over the past several decades. The regions of China believed to be most sensitive to ecosystem damage from acid rain are the subtropical evergreen forests of southern China, the high-altitude tundra areas of Tibet and parts of Qinghai Province, the coniferous forests in northeastern China, and areas of southwestern and eastern China that are expected to have extremely high deposition in the future.

Based on the RAINS-Asia model, sulfur deposition already exceeds the critical load (box 2.3) in Sichuan and parts of eastern China and is expected to increase there and elsewhere in the near future (map 1). In southwestern China, especially around Chongqing, annual depositions already exceed the critical load by as much as 11,000 milligrams per square meter. Some of the most sensitive ecosystems in southern China may already have been damaged by acidic precipitation. Although considerable research needs to be done on the direct impact of sulfur on rice, studies suggest that damage begins to occur on similar ecosystems at sulfur dioxide ambient concentration threshold values of 20–30 $\mu g/m^3$ (Downing, Ramankutty, and Shah 1997). During the winter average ambient sulfur dioxide concentrations are above 600 $\mu g/m^3$ in Chongqing. A survey of forest growth in Sichuan Basin and Guizhou Province has found that forest damage (indicated by rates of tree and foliage growth) is higher in areas with highly acidic rain. A 50 percent drop in biomass production in commercial spruce forests experiencing acid rain has also been reported (Zhao and Xiang 1988). Areas receiving acid rainfall also show higher rates of material corrosion in exposure tests.

In 1995 the Chongqing Environmental Protection Bureau carried out the most detailed analysis to date of crop and forest loss due to acid rain. An assessment of vegetable and crop losses was carried out for the nine central districts of Chongqing Municipality. Ambient concentrations of sulfur dioxide in these districts averaged 390 $\mu g/m^3$. The RAINS-Asia model projects that annual acid deposition in central Chongqing is on the order of 12,356 milligrams a square meter, while excess deposition is about 11,379 milligrams a square meter. Surveys by the Chongqing Environmental Protection Bureau found that about 24 percent of the vegetable

crop was damaged by acid rain in 1993, amounting to a loss of about 62 million yuan. Similar losses in production were found for grain crops, with losses totaling about 184 million yuan. Total losses for the forestry sector (both for trees that have died and for reduced growth) were estimated at 169 million yuan.

Ambient concentration data from Chinese sources and deposition estimates from the RAINS-Asia model put estimated current crop and forest losses at $5 billion a year. The average crop loss for provinces experiencing acid rain was estimated at 3.6 percent.[5]

Climate change

The increasing concentration of carbon dioxide in the atmosphere as a result of fossil fuel combustion and other human activities is known as the *greenhouse effect*. The greenhouse effect can cause changes in the earth's climate, resulting in damage to agricultural production, terrestrial and aquatic ecosystems, human settlements, and human health. Based on current trends in greenhouse gas emissions, leading climate change models project an increase in earth's mean surface temperature of 1.0–3.5 degrees Celsius (°C) by 2100 and an associated increase in sea level of about 15–95 centimeters (Watson and others 1996). While the science behind these models is improving, substantial uncertainty remains about the climate change process and impact predictions, particularly for individual regions.

Chinese researchers predict that increased greenhouse gas concentrations will lower rice, wheat, and cotton production because of higher temperatures, increased soil evaporation, and more frequent and severe storms. Rising sea level is another concern for China and other countries with large populations living in low-lying areas. According to Chinese studies, a 1 meter rise in sea level, combined with storm surges and the tide, could flood an area of China the size of Portugal (including the cities of Shanghai and Guangzhou), displacing 67 million people at current population levels (Johnson and others 1996).

Although the risks from climate change are substantial, this report makes no attempt to quantify the associated economic losses. Aside from great uncertainty about potential damages and the complications of valuation (see below), most of the effects of climate change

---

BOX 2.3

What is critical load?

The concept of critical load has been used in European acid rain negotiations and in the analysis of acid rain impacts conducted for the RAINS-Asia model (Downing, Ramankutty, and Shah 1997). Critical load is defined as the highest deposition of acidic compounds that can occur without causing harmful long-term effects on ecosystem structures and functions. Areas that have a limited natural capacity to absorb or neutralize acid rain have a low critical load. Ecosystems that are better able to buffer acidity (through different soil chemistry, biological tolerances, or other factors) have a higher critical load. Assessing the natural capacity of ecosystems to withstand levels of pollution is a method of measuring ecosystem health and of assessing the environmental benefits of emission reductions.

Source: World Bank staff.

are expected to occur beyond the time frame of this study (that is, after 2020). Local and regional pollution is expected to cause more damage in the short term.

Economic valuation of impacts

Valuation of the physical damages of air and water pollution can provide a useful guide to environmental policy. For individual pollutants it is possible to find the least-cost means of abatement without assigning monetary values. One way is to rank the costs of various options for meeting environmental standards or other targets. But when policymakers have to compare the costs of different pollutants—to human health or productivity—cost-effectiveness analysis based on physical damages is not feasible. How should a municipality decide how much of its budget to devote to controlling air pollution relative to water pollution, for example? Or, within the water pollution budget, how much should be devoted to controlling biological oxygen demand pollution and how much for heavy metals?

The most common way to create a comparative framework is to convert physical damages to economic values. The economic valuation of a human life poses the greatest difficulty because it raises social, political, and ethical questions. To value the excess mortality attributable to pollution, this report uses measures of willingness to pay, the standard methodology for valuing environmental impacts in industrial countries. A mid-range estimate of the willingness to pay to avoid a premature death in the United States is about $3 million per statistical life. This estimate can be conservatively scaled for use in China by multiplying by the ratio of average per capita incomes in the two countries ($500/$25,000), which results in an estimate of $60,000 per statistical life in urban areas and $31,800 in rural areas. Much lower values are obtained by using the "human capital" approach, which includes only the lost wages due to a premature death. Lost wages total about $9,000 in urban areas and $4,800 in rural areas.[6] Premature deaths caused by air pollution cost about $14 billion a year under the willingness-to-pay valuation and $2 billion a year under the human capital valuation (table 2.4). However, disease and illness account for far more damages under both the willingness-to-pay and human capital approaches.

TABLE 2.4

Air and water pollution cost China about 8 percent of GDP
(millions of U.S. dollars)

Problem	Willingness-to-pay valuation	Human capital valuation
Urban air pollution	32,343	11,271
Premature deaths	10,684	1,597
Morbidity	21,659	9,674
Restricted activity days	3,842	3,842
Chronic bronchitis	14,092	2,107
Other health effects	3,725	3,725
Indoor air pollution	10,648	3,711
Premature deaths	3,517	526
Morbidity	7,131	3,185
Lead exposure (children)	1,622	270
Water pollution	3,930	3,930
Health care costs	1,988	1,988
Agricultural and fishery losses	1,159	1,159
Water shortages	783	783
Acid rain	5,046	5,046
Crop and forest damage	4,364	4,364
Materials damage	271	271
Ecosystem damage	411	411
Total	53,589	24,228
Percentage of GDP	7.7	3.5

Source: Annexes 2.1 and 2.2.

The valuation of other impacts, such as the medical costs of illness or damage to agricultural crops, is not as contentious. The health and productivity losses associated with urban air pollution, including hospital and emergency room visits, lost work days, and the debilitating effects of chronic bronchitis, are estimated at more than $20 billion a year, making them the single largest environmental cost in China today. Although comprehensive data on lead poisoning in China are not available, it is possible to estimate the lead exposure of children based on numerous studies of lead poisoning (annex 2.1).

Total air and water pollution costs are conservatively estimated at $54 billion a year, or roughly 8 percent of GDP (see table 2.4). The largest losses to health, productivity, and agriculture are:

• Health losses associated with urban air pollution—particularly debilitating chronic bronchitis.

• Health losses associated with indoor air pollution.

• Chronic disease from water pollution—especially heavy metals and toxics.

• Crop and forestry damage from acid rain.

• Nervous system damage and reduced intelligence among children exposed to high levels of lead.

These losses vary among provinces because each has a different balance of environmental problems. For example, water shortages and attendant pollution are more pervasive in northern China, while sulfur contamination and acid rain are more of a problem south of the Yangtze River. Nonetheless, with total losses of about $54 billion a year, national policymakers cannot help but be concerned about China's environment.

Conclusion

As Premier Li Peng indicated in July 1996, improving China's environment is essential to realizing the potential improvements in the quality of life that rising incomes should bring. Today the damages of excessive pollution—in the form of premature deaths, sickness, and damage to productive resources and urban infrastructure—are estimated to cost the Chinese about 8 percent of GDP.

These costs are high. But economic reforms and environmental policies have held them in check. Will this continue? That is the subject of the next chapter.

Annex 2.1 Economic valuation of air and water pollution damage

This annex describes the methodology used to estimate the environmental damage associated with air and water pollution in China, and to value these damages in monetary terms. The four types of environmental costs outlined here are urban air pollution, indoor air pollution, water pollution, and lead pollution. Acid rain damage and cost estimates are provided in annex 2.2.

Urban air pollution

Damages from urban air pollution were estimated using dose-response functions from Chinese and international studies. Epidemiological studies carried out in China, including some supported by the World Health Organization (WHO) and the World Bank (Xu 1993), were used to assess the relationship between ambient air pollution (concentrations of total suspended particulates and sulfur dioxide) and several morbidity endpoints (hospital admissions and emergency room visits) and mortality. For other morbidity estimates—asthma attacks, lower respiratory infections, respiratory symptoms, restricted activity days—international studies were used. International dose-response functions for health effects were used where they had not been directly measured in China. The dose-response functions used to estimate mortality and morbidity effects from air pollution are shown in annex 2.1 table 1 (these same dose-response functions were used for indoor air pollution; see below).

Studies in China have found a statistically significant relationship between health outcomes and both particulates (total suspended particulates and PM-10) and sulfur (sulfur dioxide). The interpretation here is that sulfur, generated primarily by coal combustion in China, is a good proxy for fine particulates, and that a portion of the fine particulates are formed from sulfur in the atmosphere. Thus we have used fine particulates (PM-10) as the key measure of health impacts in this report. Because only total suspended particulates are measured consistently in China, PM-10 has been calculated throughout as 1 microgram per cubic meter ($\mu g/m^3$) of total suspended particulates=0.6 $\mu g/m^3$ of PM-10.

The valuation of physical impacts is based on two approaches: willingness-to-pay and human capital. As the field of environmental economics has developed, the preferred approach to valuing environmental damages to health has shifted from the human capital approach to the willingness-to-pay approach. The human capital approach values mortality and morbidity impacts simply as lost productivity (discounted lost wages) plus out-of-pocket expenditures (such as medical bills). In essence, the valuation of human life or health under this approach is similar to determining the value of any physical asset (Hara, Anis, and Stieb 1995). The human capital approach continues to be used because it is a

ANNEX 2.1 TABLE 1
Dose-response functions for PM-10

Health impacts	Unit[a]
Mortality (deaths)	6
Respiratory hospital admissions (cases)	12
Emergency room visits (cases)	235
Restricted activity days (days)	57,500
Lower respiratory infection/child asthma (cases)	23
Asthma attacks (cases)	2,608
Chronic bronchitis (cases)	61
Respiratory symptoms (cases)	183,000

a. Number of additional deaths, cases, or days per 1 million people for every 1 $\mu g/m^3$ increase in ambient concentrations of PM-10.
Source: World Bank staff estimates.

convenient way to estimate a complicated phenomenon, and because wages are easily observed.

The willingness-to-pay approach measures the value of human life and health in the marketplace (Cropper and Oates 1992). What are individuals willing to pay to reduce the risk of injury or death, or willing to accept for taking on a limited risk? In properly functioning markets, where risks are known, there is likely to be a risk premium, relative to jobs requiring similar skills, for risky occupations. The amount people pay to reduce the risk of injury or death also can be observed (home fire alarms, seatbelts, safety seats, air bags), as can the amount people are willing to accept to avoid limited risk (life, health, and flight insurance). Given the current transition from a planned to a market economy, markets for risk in China are still in their infancy, and no known willingness-to-pay survey has been carried out. To provide internationally comparable estimates of premature mortality costs, this report estimated the willingness to pay to avoid premature mortality in China. These values have been converted from U.S. estimates comparing wage rates for risky and nonrisky occupations and from other contingent valuation studies that estimate the value of making small reductions in the probability of death (annex 2.1 table 2).

For the human capital estimates, age-specific mortality data for Beijing (1989) were used to calculate the number of "excess" deaths that could be avoided for each age group by meeting Chinese class 2 air pollution standards. The number of life-years remaining for each age group has been calculated using age-specific life expectancy tables for China. Thus if a child died before age 1, the number of life years lost is 68, which is life expectancy at birth. Average life-years remaining for each age group are multiplied by the number of deaths avoided to obtain the number of life-years saved. Average annual urban ($866) and rural ($289) wages were used to value the number of avoided mortalities (human capital approach) and morbidities from assumed improvements in ambient urban air quality and rural indoor air quality, respectively. For urban wages an additional 30 percent was added to account for remaining nonwage benefits such as housing, schooling, and pensions.

Indoor air pollution

While many urban residents are exposed to indoor air pollution, estimates of indoor air pollution here are limited to the rural population to avoid double counting. Starting with the rural population of 859 million in 1995 (438 million men and 421 million women), it was assumed that:

• Only the half of the population living in temperate climates is exposed.

• Exposure in temperate areas is limited to the three winter months, when additional fuels are used for heating and doors and windows are kept closed.

• Rural women spend as much time indoors as their urban counterparts (in temperate areas during winter), but rural men spend only two-thirds as much time indoors as their urban counterparts.

This leaves an exposed population of 109 million (44 million men and 65 million women), or 13 percent of the rural population.

Fuel use data for rural areas were used to calculate the percentage of the population exposed to coal and to

ANNEX 2.1 TABLE 2
Unit valuation of mortality and morbidity
(U.S. dollars)

Health impact	Urban areas (ambient air)		Rural areas (indoor air)	
	Willingness-to-pay approach	Human capital approach	Willingness-to-pay approach	Human capital approach
Mortality, per death	60,000	8,970	31,762	4,748
Respiratory hospital admissions, per case	284	284	150	150
Emergency room visits, per case	23	23	12	12
Restricted activity days, per day	2.32	2.32	1.23	1.23
Lower respiratory infections/child asthma, per case	13	13	7	7
Asthma attacks, per case	4	4	2	2
Chronic bronchitis, per case	8,000	1,196	4,235	633
Respiratory symptoms, per case	0.60	0.60	0.32	0.32

Source: World Bank staff estimates.

biomass fuels. These figures were then used to calculate the average concentrations of total suspended particulates inside rural households, using concentration data from Sinton and others (1995). "Excess" pollution was then calculated comparing average indoor concentrations to Chinese standards. From the excess pollution data the physical damages were calculated in the same way as for urban air pollution.

Rural wages were used to calculate indoor air pollution losses using an approach similar to that for urban air pollution (see annex 2.1 table 2).

Water pollution

A lack of data prohibited an independent estimation of water impacts for this report. Instead, estimates prepared by Chinese researchers as background for this report were used. The Chinese researchers used studies carried out in areas near Shenyang and Fushun where land had been irrigated with highly polluted water, and estimated the resulting impacts on agricultural production. The Chinese researchers also compared disease rates among fishermen and farmers living near the Bohai and Huanghai seas and among persons in the same county in Northeast China who were near and far from the Songhua River. In both studies the subjects living near the water bodies had significantly higher rates of liver cancer and other water-related diseases. Studies conducted by researchers from Harvard University and China in Anhui Province also found much higher rates of spontaneous abortion among people dependent on pond water than on well water. Despite the difficulty of generalizing the results of these limited studies, it was conservatively estimated that the health costs of water pollution are at least $2 billion a year.

ANNEX 2.1 TABLE 3
Costs of lead exposure for children
(millions of U.S. dollars)

Cost	United States (per 1 µg/dl)	China (per 1 µg/dl)	Excess (8 µg/dl)
Medical costs	189	5.5	44.2
Compensatory education	481	14.0	112.5
Earnings	5,060	147.9	1,183.3
Infant mortality	1,140	33.3	266.6
Neonatal care	67	2.0	15.7
Total	6,937	202.8	1,622.2

Source: Schwartz 1993; World Bank staff estimates.

Lead pollution

U.S. studies have estimated the benefit of reducing average blood-lead levels by 1 microgram per deciliter (µg/dl) in both children and adults (Schwartz 1993). To estimate the total benefits of reducing lead pollution, unit costs (1 µg/dl) were multiplied by the amount that the average blood-lead level exceeds the threshold for lead impacts; the current definition of lead poisoning in the United States is more than 10 µg/dl of blood. For China the costs of lead pollution for children were calculated by scaling the U.S. figures by per capita incomes in each country ($500/$25,000) and by the comparative number of children under six living in urban areas. To obtain total costs, however, the average blood-lead level of children in Chinese cities is needed. The lack of comprehensive studies on blood-lead levels in Chinese cities precludes accurate estimates of average blood-lead levels. Still, a number of studies of lead in China have been carried out (Shen and others 1996). The estimates presented in this report assume that the average blood-lead level in urban areas of China exceeds the impact threshold by at least 8 µg/dl (annex 2.1 table 3). Other estimates of lead impacts can be obtained by multiplying alternative excess blood-lead levels by the per unit figures presented in the second column of annex 2.1 table 3.

Annex 2.2 Simulating acid rain impacts using the RAINS-Asia model

The acid rain scenarios in the report were developed using the Regional Air Pollution Information and Simulation Model-Asia (RAINS-Asia) model, a tool for integrated analysis of air pollution. The model was created through an international collaborative program designed to study the effects of sulfur deposition in Asia. It is based on the framework of a pioneering acid rain prototype for Europe (RAINS-Europe) that is being used for policy analysis under the European Convention on Long-Range Transboundary Air Pollution.

The RAINS-Asia model covers East, South, and South-East Asia, with particular emphasis on China, India, Indonesia, Japan, the Republic of Korea, and Thailand. The model contains databases on energy consumption for 23 countries, 94 regions, and 250 large point sources. The model also estimates the carrying capacity for acid deposition of 31 types of ecosystems. Values for sulfur depositions are based on a yearly average and are calculated at a 1 x 1 degree resolution. Aggregate results of source and receptor information on acid deposition patterns for each region or country can be obtained. Local impacts of acid deposition can be estimated as well. The temporal range of the model is 1990–2020. A detailed description of the model and its uses can be found in Downing, Ramankutty, and Shah (1997).

Acid deposition can have a range of impacts, including:
• Acidification of soils and damage to trees and other vegetation, including agricultural crops and forests.
• Acidification of lakes and rivers, leading to fish loss.
• Eutrophication of surface waters due to excess nitrogen inputs.
• Direct impacts on vegetation due to exposure to sulfur dioxide and nitrogen oxides in the vicinity of major emitting sources.
• Corrosion of metals and building materials.
• Health effects caused by increased levels of sulfate and other aerosols.

In Asia a limited amount of work has been conducted on acid rain impacts and their costs. Studies of acid rain impacts have been carried out in parts of southwestern China. Some of the impacts seen in the vicinity of Chongqing, Sichuan, and Guiyang, Guizhou, include 50 percent dieback of forests on Nanshan mountain, more than 50 percent reduction in biomass production in commercial spruce forests experiencing acid rain with a pH of less than 4.5, yellowing of rice in large areas near Chongqing after rainfall with a pH of less than 4.5, and wheat fields that are seriously affected by rainfall with a pH of less than 3.5 during the middle and late growing periods.

Simulations with the model show that exceedences—that is, deposition of sulfur exceeding the critical loads of the ecosystem—are most severe in and around urban areas such as Shanghai and large regions around Chongqing (see map 1). It is estimated that nearly 10 percent of the total land area is exposed to more than 1,000 milligrams per square meter (mg/m^2) a year of sulfur deposition in excess of critical loads. About 1 percent of the total land area (concentrated in and around Chongqing) has exceedences of

ANNEX 2.2 TABLE 1
Estimates of sulfur dioxide damage to farming and forestry, 1995

Province	Gross output (billions of yuan)			Annual sulfur dioxide deposition (mg/m^2)	Ambient sulfur dioxide concentrations	Share of output damaged (percent)	Total damage $(10^8$ yuan)
	Farming	Forestry	Total				
Liaoning	392	13	405	1,500	50	1.9	7.70
Beijing	87	3	90	1,700	56	2.2	1.98
Tianjin	85	1	86	3,100	104	5.2	9.15
Hebei	754	24	778	1,500	50	2.0	15.56
Shanxi	203	13	216	1,000	33	0.8	1.73
Shanghai	78	0	78	4,600	154	8.5	6.63
Jiangsu	986	21	1,007	4,200	141	7.6	76.53
Zhejiang	482	50	532	2,500	84	4.0	21.28
Anhui	638	40	678	2,100	71	3.2	21.70
Fujian	340	59	399	1,300	44	1.5	5.99
Jiangxi	332	41	373	1,800	60	2.5	9.33
Shandong	932	42	974	2,300	77	3.8	37.01
Henan	866	38	904	2,000	66	3.0	27.12
Hubei	612	28	640	1,800	60	2.5	16.00
Hunan	580	42	622	1,500	50	1.9	11.82
Guangdong	778	46	824	1,900	62	2.6	21.42
Guangxi	384	33	417	1,600	53	2.0	8.34
Sichuan	887	45	932	2,900	97	5.0	46.60
Guizhou	224	15	239	3,000	101	5.1	12.19
Yunnan	299	41	340	400	12	0.0	0.00
Shaanxi	258	17	275	1,300	43	1.5	4.13
Total	10,197	612	10,809	44,000			362.19 ($4.36 billion)

Source: World Bank staff estimates.

more than 5,000 mg/m² a year of sulfur deposition. Chongqing has depositions that exceed 11,000 mg/m² a year.

Studies conducted in Chongqing provide information to arrive at a stylized relationship between deposition, ambient exposure, and crop loss. An exponential dose-response function was derived linking some data points available from field studies in Chongqing. A similar dose-response function was assumed for impacts on forestry. The dose-response function was applied to deposition data available from the RAINS-Asia deposition maps and from farming and forestry output to obtain damage results (annex 2.2. table 1). The resulting estimate of economic costs from crop and forestry losses alone is more than $4 billion a year. This is a conservative estimate, since it does not include impacts on materials or human health.

Notes

1. Throughout this chapter, the disease burden in China is measured in terms of disability-adjusted life years, or DALYs. For an explanation of DALYs and the disease burden in China, see World Bank (1996b), annex 3.

2. See the research by X. Xu and colleagues, a partial list of which is given in Wells, Xu, and Johnson (1996).

3. Reductions in morbidity from better water supply and sanitation are 26 percent for diarrhea, 27 percent for trachoma, 29 percent for ascariasis, 77 percent for schistosomiasis, and 78 percent for dracunculiasis. Mean reductions in diarrheal-specific mortality are 65 percent, while overall child mortality drops by 55 percent (Esrey and others 1991).

4. The RAINS-Asia model estimates sulfur emissions from two sources. Large point sources, of which there are 161 in China, are large power plants. All other sources, including small power stations, industrial boilers, residential fuel use, and transportation, are considered "area" sources.

5. The empirical evidence for these estimates is limited to a few Chinese studies. The uncertainty of crop yield estimates are further aggravated by other confounding variables, such as ozone and other pollutants (annex 2.2).

6. The human capital approach is the most common methodology used in China. A typical value of a statistical life used by Chinese social scientists is 70,000 yuan, or about $8,400.

References

Cropper, M.L., and W.E. Oates. 1992. "Environmental Economics: A Survey." *Journal of Economic Literature* 30 (June): 675–740.

Delft Hydraulics, Mott MacDonald, and Welsh Water International. 1995. "Hangzhou Bay Environmental Study." Report prepared for the World Bank's International Development Association. Washington, D.C.

Downing, T., R. Ramankutty, and J. Shah. 1997. "RAINS-Asia: Regional Air Pollution Information and Simulation Model." World Bank, Washington, D.C.

Esrey, S.A, J.B. Potash, L. Roberts, and C. Shiff. 1991. "Effects of Improved Water Supply and Sanitation on Ascariasis, Diarrhoea, Dracunculiasis, Hookworm Infection, Schistosomiasis, and Trachoma." *Bulletin of the World Health Organization* 69 (5): 609–21.

Fisher, A., L.G. Chestnut, and D.M. Violette. 1989. "The Value of Reducing Risks of Death: A Note on New Evidence." *Journal of Policy Analysis Management* 8: 88–100.

Hara, D., A. Anis, and D. Stieb. 1995. "Review of Methods for Valuing Human Injury and Lost Productivity Under Changes in Air Quality." Health Canada, Ottawa, Canada.

Johnson, T.M., J.F. Li, Z.X. Jiang, and R.P. Taylor, eds. 1996. *China: Issues and Options in Greenhouse Gas Emissions Control.* World Bank Discussion Paper 330. Washington, D.C.

Pope, C.A., and X. Xu. 1993. "Passive Cigarette Smoking, Coal Heating, and Respiratory Symptoms of Nonsmoking Women in China." *Environmental Health Perspectives* 101(4): 314–16.

Schwartz, Joel. 1993. "Societal Benefits of Reducing Lead Exposure." *Environmental Research* 66: 105–24.

Shen, X.M., J.F. Rosen, D. Guo, and S.M. Wu. 1996. "Childhood Lead Poisoning in Children." *Science of the Total Environment* 181: 1001–109.

Sinton, Jonathan E., Kirk R. Smith, Hansheng Hu, and Junzhuo Liu. 1995. *Indoor Air Pollution Database for China.* Environmental and Occupational Epidemiology Series. World Health Organization, Global Environmental Epidemiology Network, Geneva.

Streets, David G. 1996. "Acid Rain in Asia." World Bank, Washington, D.C.

Watson, R. T., M.C. Zinyowera, R.H. Moss, and D.J. Dokken, eds. 1996. Climate Change 1995: Impacts, Adaptations and Mitigation of Climate Change: Scientific-Technical Analyses. Intergovernmental Panel on Climate Change. Cambridge: Cambridge University Press.

Wells, G.J., X.P. Xu, and T.M. Johnson. 1994. "Valuing the Health Effects of Air Pollution: Application to Industrial Energy Efficiency Projects in China." World Bank, Washington, D.C.

World Bank. 1992. "China: Environmental Strategy Paper." China and Mongolia Department, Washington, D.C.

———. 1993. *World Development Report 1993: Investing in Health.* New York: Oxford University Press.

———. 1996a. "China: Chongqing Industrial Pollution Control and Reform Project." Staff Appraisal Report. Washington, D.C.

———. 1996b. "China: Issues and Options in Health Financing." China and Mongolia Department, Washington, D.C.

Xu, X., B.L. Li, and H.Y. Huang. Forthcoming. "Air Pollution and Unscheduled Hospital Outpatient and Emergency Room Visits." *Environmental Health Perspectives.*

Xu, X., J. Gao, D.W. Dockery, and Y. Chen. 1994. "Air Pollution and Daily Mortality in Residential Area of Beijing, China." *Archives of Environmental Health* 49(4): 216–22.

Xu, Z.Y. 1993. "The Report of the WHO Consultation on Air Pollution and Health." World Bank, China and Mongolia Department, Washington, D.C.

Zhao, D., and Y. Xiang. 1988. "Acid Rain in Southwestern China." *Atmospheric Environment* 22(2): 349–58.

chapter three

Pollution, Health, and Productivity: Plotting Alternative Futures

China has set ambitious quantitative objectives for reducing air and water pollution by 2000, as well as more qualitative objectives in its long-range plan to 2010. The quantitative goals include containing the increase in wastewater discharges to 14 percent between 1995 and 2000 and holding particulate and sulfur dioxide emissions nearly constant (box 3.1). For the qualitative goals, by 2010 the government hopes to have reversed increasing environmental deterioration and ecosystem degradation. Drinking water in major cities will be expected to meet national standards and surface water to meet standards for appropriate use. Urban air should meet class 2 standards. Finally, the government plans to significantly reduce discharges of major industrial pollutants, halt the deterioration of major river basins, and bring an end to worsening acid rain.

Although past economic growth has generated high environmental costs, future growth will give China the opportunity to correct the damage. With continued rapid growth newer and cleaner technologies can replace outmoded and dirty ones. And with rising incomes Chinese consumers will buy different products and services than in the past, propelling changes in the structure of China's economy. A cleaner, safer future is not a given, however.

Will rapid growth, combined with existing policies, be sufficient to achieve China's ambitious environmental objectives? If not, could a different pattern of investment allow China to both grow rapidly and achieve its environmental objectives? This chapter answers these questions by exploring three alternative scenarios. The first, "business as usual" scenario examines the beneficial effects of projected technological and structural change on environmental quality based on the assumption that current environmental policies remain broadly unchanged. Although technological and structural change will bring benefits, the business as usual scenario will not be sufficient to change the pattern of growth and achieve China's ambitious environmental objectives. Thus two alternative scenarios are presented to show the changes in policies and investment patterns China will require to achieve its goals.

Basic assumptions about macroeconomic and sectoral growth

During 1995–2020 real GDP growth is expected to average 6.7 percent a year.[1] Two factors inherent to growth have important consequences for the environment. The first is the rate of investment, assumed here to gradually fall from its high current rate of more than 40 percent to 34 percent by 2020. The capital stock accumulated as a result of new investment generally acquires the characteristics of advanced technology available elsewhere. This is an optimistic assumption for China, however, because it presumes that domestically manufactured capital equipment is comparable in abatement technology to equipment manufactured elsewhere. If the economy continues to grow rapidly, the contribution of the current capital stock to future output will be quite small after 2005. As this capital stock is replaced by new technology, technical progress and

BOX 3.1

China's ambitious objectives

Many governments commit themselves to only vague environmental objectives; China, by contrast, has defined a clear set of measurable targets. Achieving the targets for lower water pollution listed in the table below will require increasing industrial wastewater treatment from 63 percent to 74 percent and combined municipal and industrial wastewater treatment from 19 percent to 25 percent. Moreover, the government plans to invest 450 billion yuan in abatement technology between 1996 and 2000, with about half going toward controlling air pollution and 40 percent being used to cut back wastewater discharges.

China's ambitious environmental targets for 2000

Pollutant	1995 level	2000 target
Water		
Wastewater discharges (billions of tons)	42.2	48
Industrial	27.1	30
Municipal	15.1	18
Chemical oxygen demand (millions of tons)	22.3	22
Industrial		16
Municipal		6
Air (millions of tons)		
Smoke dust	17.4	17.5
Industrial fugitive dust	17.3	17.0
Sulfur dioxide	23.7	24.6
Solid waste (millions of tons)		1,110
Industrial		930
Municipal		180
Industrial solid waste disposal (millions of tons)	62	60

Source: NEPA, SPC, and SETC 1996.

stricter environmental controls should lower both emission coefficients and total pollution.

A second consequence of growth is the structural shifts it induces in the composition of demand. Over time the share of agriculture in China's GDP will undoubtedly fall and the share of services will increase, while the shares of industry and construction will remain roughly constant. Since services tend to be less polluting than agriculture, this shift will also likely cause pollutant emissions to fall.

Demand for energy will also be affected by growth and will rise over the next twenty-five years. Demand for primary energy in thermal power generation and heat production is expected to reach 1,430 million tons of coal equivalent (Mtce) by 2020, while demand for primary electricity production in nonthermal power plants (from hydro, nuclear, and others) will rise to 93 Mtce.[2]

Energy prices will have an important effect on energy intensities (that is, energy consumed per unit of GDP). Low average prices for coal—perhaps 10–20 percent below the equivalent border parity price during 1990–95[3]—have encouraged its inefficient use. Further, the environmental damage caused by coal use is worsened by the high ash content of much of the coal that is delivered and the inconsistent quality of different grades of coal. Although most petroleum prices are close to their equivalent border parity prices, low taxes on gasoline and diesel fuel could be raised to discourage use.[4]

Assuming that these relative prices do not change, primary energy demand is projected to increase from about 1,240 Mtce in 1995 to about 3,700 Mtce in 2020 (figure 3.1). Final energy demand will grow from about 920 Mtce to 2,630 Mtce over the same period (figure 3.2).[5] Coal accounted for 78 percent of primary energy demand in 1995; this share is projected to decline to 70 percent by 2020. However, coal use will shift from small and medium-scale users to large users such as power plants and large heating boilers. Final use of coal will decline from 60 percent of final energy consumption in 1995 to 40 percent in 2020. This shift could bring enormous environmental benefits because it is much easier to control and monitor emissions from large plants than from myriad small and medium-scale users.

Taken together, changing consumption patterns and more efficient technologies could easily reduce China's energy intensity to 62 percent of its 1995 level by 2020.

This drop will be particularly sharp over the next ten years, with energy intensity falling to 80 percent of its 1995 level by 2005. In 1995 China had an energy consumption to GDP ratio of 1.8 kilograms of coal equivalent per U.S. dollar (similar to India's). By 2020 that ratio should improve to 1.1. Even then, however, it would be far above current values for Japan (0.2) and the United States (0.4).

Pollution without policy changes: The business as usual scenario

The central idea of the business as usual scenario is that it allows for those improvements in environmental performance that accompany market-driven changes in the economy.[6] Under this scenario there will be no significant change in environmental policies or in the degree of regulatory pressure used to enforce those policies. But even in the absence of a concerted effort to address environmental problems, the advent of cleaner production techniques, the redesign of existing technologies, the shift toward higher value-added products and activities, and other such changes will cause emission coefficients to fall as China grows.[7]

Bearing these considerations in mind, we address two questions in the following sections: How will economic growth affect emissions? And how will emissions affect air and water quality?

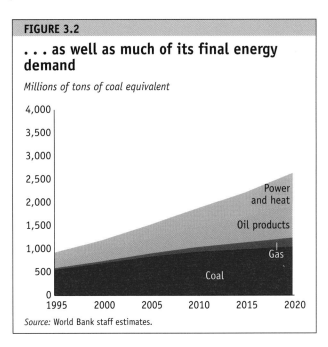

FIGURE 3.1

Coal will continue to provide most of China's primary energy demand . . .

Millions of tons of coal equivalent

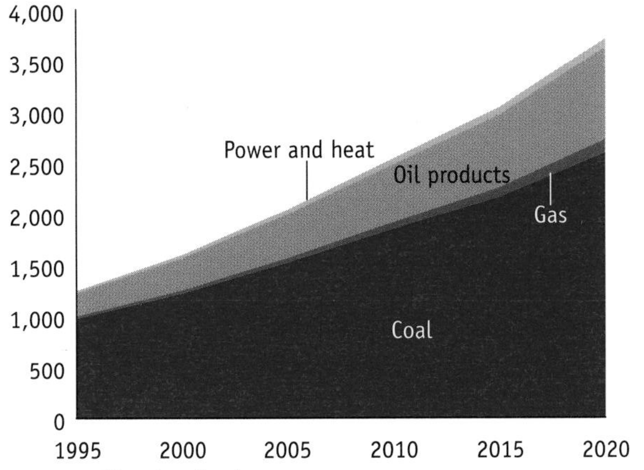

Power and heat

Oil products

Gas

Coal

Source: World Bank staff estimates.

FIGURE 3.2

. . . as well as much of its final energy demand

Millions of tons of coal equivalent

Power and heat

Oil products

Gas

Coal

Source: World Bank staff estimates.

Future pollution loads

Particulates. Fine particulate emissions will increase by 39 percent between 1995 and 2010 but will decline slowly thereafter (figure 3.3). By 2020 particulate emissions will be 35 percent higher than in 1995. This outcome represents the sum of two sets of forces. Better particulate controls for large sources—heavy industry and power and heat plants—will reduce emissions as older equipment is replaced, causing their share in total emissions to fall from 60 percent in 1995 to 52 percent in 2020. Emissions in the residential, commercial, and small industrial sectors will increase, however. New technology offers much smaller gains in these sectors—except to the extent that more coal washing, improved energy efficiency, and better boiler and stove designs lower average emissions per ton of coal burned.

Unless policies are changed, there will be substantial growth in fine particulates emitted from small sources. Pollution discharged from low chimneys is the main contributor to poor air quality in urban areas. Under the business as usual scenario the amount of smoke emitted by households will grow by 90 percent between 1995 and 2020. Thus particulate emissions would stabilize at a level that is associated with extremely poor air quality for most urban households, which continue to rely on coal as the primary fuel for heating and cooking.

Sulfur dioxide. Under the business as usual scenario sulfur dioxide emissions will increase 150 percent by 2020 (figure 3.4). As with particulates, growth in the use of coal by households and other small sources drives the initial increase in emissions. Later, an increase in demand for electricity outweighs the initial drop in sulfur dioxide emissions per ton of coal burned in power and heat plants (a drop made possible by improvements in the thermal efficiency of new power plants and by the adoption of low-cost control measures). As a result power and heat plants account for all of the growth in sulfur dioxide emissions after 2005.

Nitrogen oxides. Nitrogen oxide emissions will increase rapidly during 1995–2020 as a result of brisk growth in emissions from small sources (figure 3.5). Much of this growth is the result of increased oil consumption by motor vehicles—motorcycles, cars, trucks, and buses. Nitrogen oxide emissions from power and heat plants follow the same pattern as for sulfur dioxide emissions but with a higher overall growth rate. Although the scenario assumes that burners with lower nitrogen oxide emissions will gradually be incorporated in new boilers, this will only partly offset the effects of rapid growth in electricity demand and production.

Lead. To be consistent with the rest of the scenario, the projections for lead do not take into account the government's recent announcement of plans to phase out lead from gasoline by 2000. As such, the scenario provides a baseline against which to measure implementation of that decision.

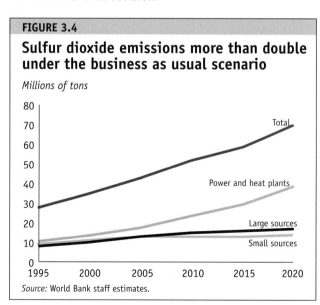

FIGURE 3.3

Particulate emissions from households, automobiles, and small industries will rise under the business as usual scenario

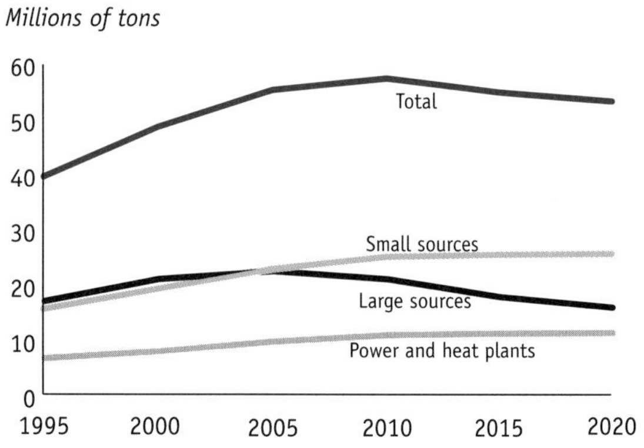

Millions of tons

Source: World Bank staff estimates.

FIGURE 3.4

Sulfur dioxide emissions more than double under the business as usual scenario

Millions of tons

Source: World Bank staff estimates.

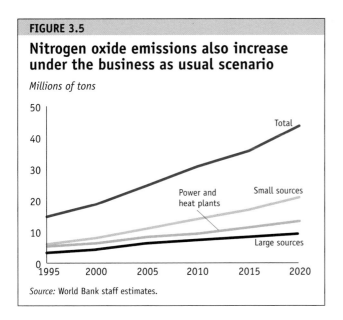

FIGURE 3.5

Nitrogen oxide emissions also increase under the business as usual scenario

Millions of tons

Source: World Bank staff estimates.

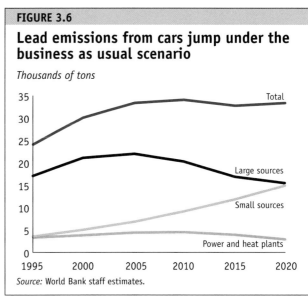

FIGURE 3.6

Lead emissions from cars jump under the business as usual scenario

Thousands of tons

Source: World Bank staff estimates.

The trend in lead emissions reflects distinct factors affecting discharges from small and large sources (figure 3.6). Better technology and controls will reduce the amount of lead emitted by large sources—nonferrous smelters, iron and steel mills, power plants—if Chinese technologies embody modern abatement technology. But increased traffic congestion and consumption of leaded gasoline will more than offset this decline, so that by 2020 lead discharges will be about 35 percent higher than in 1995. Under the business as usual scenario the average lead content of gasoline will drop by half over the period. Still, lead emissions from small sources will more than quadruple by 2020, reflecting rapid growth in the motor vehicle fleet and traffic volumes over the period—and illustrating the importance of swiftly implementing the decision to phase out lead.

Volatile organic compounds. Similar factors lie behind the quadrupling in emissions of volatile organic compounds by 2020 (figure 3.7). There is no zero- or low-cost technology—apart from simple improvements in fuel efficiency—that will substantially reduce volatile organic compound emissions per ton of petroleum products used by motor vehicles. Thus any increase in the amount of motor fuels consumed translates directly into higher emissions of these compounds.

Biological oxygen demand, suspended solids, and phosphorus. These water pollutants are considered together because they provide an indication of the

trends in industrial and nonindustrial discharges of water pollutants. No attempt has been made to estimate discharges of phosphorus from agricultural runoff, so phosphorus discharges under the business as usual scenario are almost entirely accounted for by sewage. On the other hand, about 35 percent of biological oxygen demand and about 60 percent of suspended solids discharged in 1995 came from industry.

In certain parts of China phosphorus emissions are an increasingly serious problem, causing the eutrophication of lakes and rivers that are sources of drinking water. Biological oxygen demand levels provide a general indicator of water quality and are often corre-

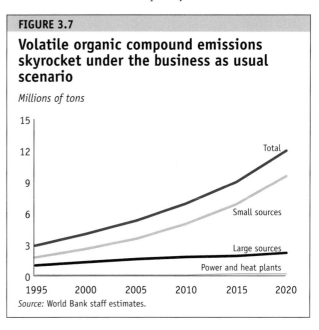

FIGURE 3.7

Volatile organic compound emissions skyrocket under the business as usual scenario

Millions of tons

Source: World Bank staff estimates.

lated with levels of coliforms, especially where untreated sewage is the main source of discharges. Where lakes and rivers are used for fish farming or industrial processes, excessive biological oxygen demand levels may impose heavy costs on enterprises. Suspended solids increase water treatment costs, and industrial discharges are often associated with phenols and other organic compounds or heavy metals whose presence may render water unfit for drinking or irrigation.

Under the business as usual scenario phosphorus emissions increase by 21 percent between 1995 and 2020. Biological oxygen demand levels rise slightly until 2005, after which improvements in industrial wastewater discharges and the spread of municipal treatment outweigh the effects of population and industrial growth (figure 3.8). However, the downward trend is eventually overtaken by population growth, and by 2020 domestic sewage accounts for 85 percent of biological oxygen demand levels. Emissions of suspended solids fall sharply after 2005 as low-cost measures to reduce industrial water pollution become more widespread (figure 3.9). Domestic sewage still accounts for 62 percent of these discharges in 2020, so measures to reduce industrial emissions will have much more impact on emissions of suspended solids than of biological oxygen demand.

Thus cleaner production technologies and better controls on industrial wastewater discharges can ensure

that overall loads of biological oxygen demand and suspended solids fall below their 1995 levels by 2020. However, further progress will depend on the extent and nature of sewage treatment. The costs may be high since the construction of sewerage networks and sewage treatment plants will be required.

Other water pollutants. Inadequate data preclude detailed projections of other industrial water pollutants. Still, trends in emissions of suspended solids, 60 percent of which originated from industry in 1995, support the view that better technologies and controls will substantially reduce industrial emissions over the next ten to twenty years, even under existing policies. Thus even stricter policies would accelerate these improvements.

Global issues. Carbon dioxide emissions under the business as usual scenario will produce 2,380 million tons of carbon in 2020, nearly three times the 800 million tons in 1995. Clearly, the international community should have a strong interest in helping China achieve its energy efficiency and diversification objectives.

Is business as usual enough?

Air quality. A simple analysis was carried out for China's largest cities (populations of more than 1 million people) and other cities (less than 1 million people)

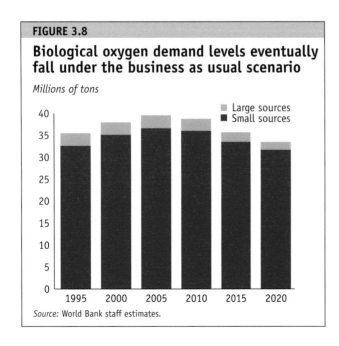

FIGURE 3.8

Biological oxygen demand levels eventually fall under the business as usual scenario

Millions of tons

Source: World Bank staff estimates.

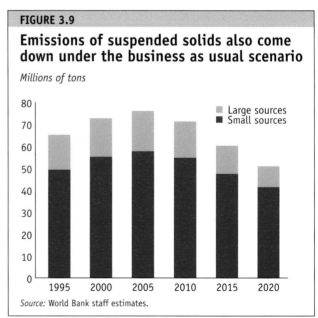

FIGURE 3.9

Emissions of suspended solids also come down under the business as usual scenario

Millions of tons

Source: World Bank staff estimates.

to examine the impact changes in emissions under the business as usual scenario will have on air quality in these areas. The average contribution of different sources of pollution (that is, power plants, large industry, and small sources, including households) was derived for eleven major cities—Beijing, Chengdu, Chongqing, Guangzhou, Harbin, Jinan, Shanghai, Shenyang, Tianjin, Wuhan, and Xian.[8] The most important determinant of air quality in China today is the quantity of coal burned by households and other small pollution sources in urban areas with populations of more than 20,000 people.[9] Thus, to the extent that overall coal consumption remains the same, the fact that one city adopts a district heating system merely represents a change in the location of pollution—not a reduction in its overall impact. If the total use of coal by households is held constant, the construction of a district heating system in Beijing or Jinan means that less coal is burned by households in these two cities but more will be burned in other urban areas. Air quality in Beijing and Jinan will be better, but air quality in other cities will be worse. Reducing the total damage caused by air pollution will require reducing the total amount of coal burned.

Historical patterns of growth in other countries suggest that large efforts may be made to improve air quality in Beijing and other large cities while it deteriorates in medium-size and small cities. Current evidence from China confirms this pattern. While China's largest cities are converting households to district heat or cleaner fuels and relocating industries to the suburbs, smaller cities lag significantly behind. At the same time, rural households and small enterprises are increasing coal use, switching from biofuels, and adopting other environmentally damaging practices.

To reflect these disparities, differentiated trends in emissions from small sources were applied to China's thirty largest cities (with a total population of nearly 70 million) and to other cities (total population of more than 230 million) and rural areas (towns and villages). In assessing the possible social costs implied by these trends, it is important to keep in mind that even the thirty largest cities will contain less than 10 percent of China's population by 2020. These emission trends were used to estimate trends in ambient air quality for the thirty largest cities (including individual estimates for eleven megacities) and other cities.

Under the business as usual scenario urban air quality will improve in the largest cities and become much worse in most other cities. However, improvements in the largest cities will not be sufficient to lower average particulate concentrations to China's own class 2 (or even class 3) ambient air quality standards. In the largest cities fine particulate concentrations will drop from 211 micrograms per cubic meter ($\mu g/m^3$) in 1995 to 126 $\mu g/m^3$ in 2020 (figure 3.10). Only three cities—Guangzhou, Shanghai, and Wuhan—will likely be in compliance with class 3 standards in 2020. In other cities fine particulate concentrations will increase from an average of 170 $\mu g/m^3$ in 1995 to 270 $\mu g/m^3$ in 2010–20. Thus tens of millions of urban residents will be exposed to fine particulate levels about four times higher than World Health Organization (WHO) guidelines. All but a small portion of fine particulate levels are accounted for by emissions from small pollution sources—households, services, commerce, and small industrial activities (figure 3.11). Thus emission control efforts will need to focus on small sources if the deteriorating trend in air quality is to be reversed.

As noted, under the business as usual scenario sulfur dioxide emissions grow much faster than fine particulate emissions—a 150 percent increase compared with a 40 percent increase. As a result average sulfur dioxide ambient levels in the largest cities will increase from 96 $\mu g/m^3$ in 1995 to 140 $\mu g/m^3$ by 2020 (figure 3.12). In other cities average ambient sulfur dioxide levels will increase

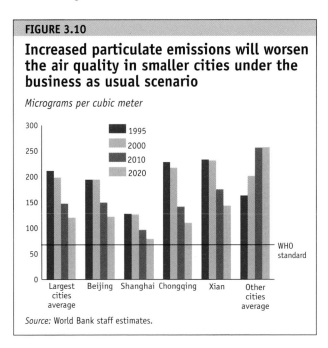

FIGURE 3.10

Increased particulate emissions will worsen the air quality in smaller cities under the business as usual scenario

Micrograms per cubic meter

Source: World Bank staff estimates.

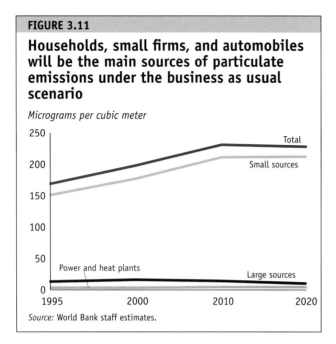

FIGURE 3.11

Households, small firms, and automobiles will be the main sources of particulate emissions under the business as usual scenario

Micrograms per cubic meter

Source: World Bank staff estimates.

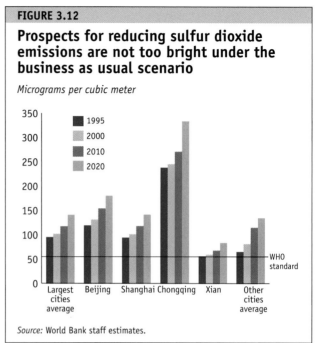

FIGURE 3.12

Prospects for reducing sulfur dioxide emissions are not too bright under the business as usual scenario

Micrograms per cubic meter

Source: World Bank staff estimates.

even faster—from 66 µg/m³ in 1995 to 135 µg/m³ by 2020. Power plants will account for most of the increase in sulfur dioxide emissions, while small sources dominate the increase in particulate emissions. Higher emissions from power plants in or near large cities do not translate into a large increase in ambient sulfur dioxide levels, however, because of the tall stacks on the power plants in these cities.[10] On the other hand, higher sulfur dioxide emissions from small sources (largely from burning huge quantities of medium- or high-sulfur coal by households and in small boilers) will result in high ambient levels and exposure to sulfur dioxide.

Regional pollution. The business as usual projections indicate that sulfur dioxide and nitrogen oxide emissions will more than double. The impact of such a jump in acid depositions on crops, forests, and surface waters will depend on the location of new power plants as well as on the capacity of the soil to buffer acid depositions.

The RAINS-Asia model was used to simulate the effects by 2020 of sulfur emissions and depositions. As map 2 shows, in more than 25 percent of the land area sulfur depositions are expected to exceed the annual carrying capacity of the ecosystems by at least 1,000 milligrams per square meter; in about 5 percent of the land area annual depositions are expected to exceed carrying capacity by 5,000 milligrams per square meter or more. Most of the excess sulfur depositions are in the rich agricultural lands in southern China.

Costs of business as usual

Estimating the current and future costs of air pollution is difficult. Reasonably reliable dose-response functions can be used to estimate the economic costs associated with damage to human health from exposure to various pollutants (box 3.2). But reducing air pollution also has other, less tangible benefits. Surveys consistently find that people are willing to pay some portion of their income to ensure that they can enjoy cleaner air.

The analysis here focuses exclusively on the costs air pollution imposes on human health. These costs are so high under the business as usual scenario that it is hardly necessary to consider the amenity and ecosystem benefits of cleaner air to justify action.

The absolute value of pollution damages in a given city for a given year depends on the reference air quality standard used in the calculations, since the gap between that standard and current exposure determines the cost. Here an annual standard of 60 µg/m³ has been used for both fine particulates and sulfur dioxide because this level corresponds to the Chinese class 2 standard and is near the upper range of WHO guidelines.

Under the business as usual scenario the health costs of exposure to particulates (for urban residents only) will rise from $32 billion (or $129 per resident exposed) in 1995 to nearly $98 billion (or $197 per resident exposed) in 2020 (figure 3.13). When expected income growth is also taken into account the costs in 2020 are some four times higher ($390 billion).

Lead exposure creates additional health care costs. Infants and children who are exposed to excessive lead suffer impairments in neurological development and a consequential loss of intelligence, reduced earnings over their lifetimes, and increased neurobehavioral problems and infant mortality. Lead poisoning among Chinese children has been estimated to cost about $1.6 billion in 1995. Under the business as usual scenario these costs will rise to more than $13 billion in 2020.

Acid rain will also become increasingly costly. If current trends persist, damage will occur to more than 25 percent of China's land, and forest and crop losses may reach $60–80 billion. Quantifying these effects, however, requires additional research.

Finally, if a "shadow price" of carbon of $20 a ton (a value currently used by the Global Environment Facility that corresponds to available estimates of marginal damage) is attached to carbon dioxide emissions, the damage from these emissions totaled $4.4 billion in 1995 and will rise to $13 billion in 2020. Although small relative to other health care damages, these costs are still significant.

Making a change: Alternative scenarios

Contrary to the assumptions underlying the business as usual scenario, policies in China are unlikely to remain static over the next two decades. But how should investments be altered to improve China's environmental future? If China began "growing differently," investing in technologies less damaging to the environment and in a less resource-intensive economic structure, emissions could fall. This section explores the most effective potential investments and quantifies their effects on the environment and on human health. Later chapters discuss the policies that would lead to their realization.

Among the changes that will reap the greatest benefits for China's environmental future are:
• Increasing substitution of cleaner fuels—especially gas and district heating—for coal for household heating and cooking than under the business as usual scenario would reduce particulate and sulfur emissions as well as greenhouse gases.
• Improving energy efficiency and conservation and diversifying energy supplies into noncoal sources would reduce ambient and solid waste emissions.
• Investing in industrial air pollution control could decrease emissions of key pollutants at least fivefold by 2020, especially with strict controls on particulates.
• Developing public transportation systems and discouraging automobile use would mitigate hydrocarbon-related air pollution in urban areas.

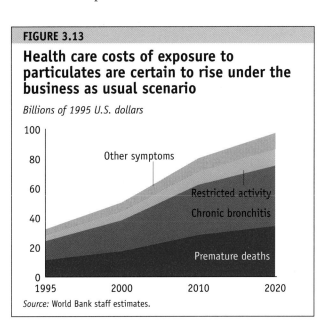

FIGURE 3.13

Health care costs of exposure to particulates are certain to rise under the business as usual scenario

Billions of 1995 U.S. dollars

Source: World Bank staff estimates.

- Raising the abatement rate for industrial water pollution and increasing the coverage and level of municipal wastewater treatment would improve water quality.

Shifting the structure of the economy in this direction is economically feasible. The speed at which the economy would adjust depends on the environmentally friendly investments China makes in the near term to avoid social costs later. Two scenarios have been examined to analyze the implications of the alternative strategies that might be followed.

Medium-cost investment scenario

A medium-cost investment scenario begins with the assumption that China will adopt policies to achieve emission standards by 2020 that are broadly equivalent to those in the United States and Europe in the early 1980s. Policy interventions (discussed in subsequent chapters) would elicit investments from both the state and nonstate sectors that would shift the economic structure along the lines suggested above. There would be more investment in cleaner fuels—including washing coal, phasing out lead more quickly, and converting from coal to gas or district heating in the residential and commercial sectors. There would also be improvements in process technologies, such as better utility and industrial boilers and the adoption of catalytic converters in new vehicles. Finally, end-of-pipe technologies would be adopted, including 95 percent particulate removal from the emissions of large industries and 99.5 percent from power plants, and primary treatment of all urban sewage and industrial wastewater.

Under the business as usual scenario investments in pollution abatement would average 0.5 percent of GDP. Under the medium-cost investment scenario significant environmental improvements could be obtained through investments averaging 1 percent of GDP. This is a 100 percent increase in environmental investment relative to the business as usual scenario. China's Ninth Five-Year Plan calls for a 70 percent increase in environmental investment over the previous planning period.[11] On an annualized cost basis—that is, annual operating costs plus the value of depreciated plants and equipment— China would invest about 2 percent of GDP in environmental protection. The state would shoulder only about 66 percent of these costs, compared with 84 percent under the business as usual scenario (see chapter 8).

High investment scenario

In many settings the marginal cost of pollution abatement rises sharply once emissions have been cut by 60–80 percent. The high investment scenario assumes that emission standards correspond to the concept of best available techniques that has been adopted by the European Union as the basis for its environmental policies.[12] The scenario assumes that by 2020 policies would induce China's commercial sector—state and nonstate—to invest in and maintain abatement levels close to what OECD countries have achieved today.

Under this scenario cleaner combustion technologies (sorbent injection, fluidized bed combustion) will become widespread in the industrial sector. End-of-pipe particulate removal will increase to 97.5 percent for large industries and to 99.8 percent for power plants. All new power plants will have flue gas desulfurization systems. Some type of particulate controls will be installed on small industrial boilers, and smokeless coal will be widely used by households and other small sources of pollution.

In the transport sector, in addition to catalytic converters for all new cars, new trucks and buses are assumed to have lean-burn engines and particulate traps. Furthermore, the scenario substantially increases public investments in infrastructure—notably gas and district heating as well as sewage treatment—to deal with emissions from dispersed sources. The scenario assumes that all urban sewage receives secondary treatment and that 80 percent of rural sewage receives primary treatment. Annual investment in pollution abatement would rise to 2 percent of GDP, and the annualized cost would rise to 4–5 percent of GDP. Because a larger portion of these costs would be borne by nonstate companies, the state's share of these costs would be about 60 percent.

Choosing a strategy

Air quality, as we have seen, has a significant impact on health. Under both the medium-cost and the high investment scenarios, particulate emissions from small sources fall to less than half their 1995 level by 2020. Average exposure to fine particulates in urban areas falls to 79 $\mu g/m^3$ under the medium-cost investment scenario and to 36 $\mu g/m^3$ under the high investment scenario (figure 3.14)—implying a dramatic reduction in premature deaths and ill health relative to the business as usual scenario. Health costs from urban air pollution are also sharply curtailed under the two alternative scenarios. By 2020 these costs fall nearly to zero in the high investment scenario (because China would achieve its own air quality standards) but exceed 10 percent of GDP in the business as usual scenario (figure 3.15).

Abating water pollution may require greater investment from both the commercial sector (state and non-state) and the public sector. Although China has managed to reduce wastewater discharges from industrial sources, industrial emissions account for only 35 percent of biological oxygen demand levels, and large industrial sources account for just one-quarter of industrial emissions. Real progress in reducing water pollution will require large investments in sewage collection and treatment. Progress under the medium-cost investment scenario will likely be sufficient to address the most pressing water pollution problems. Going beyond this program would require more careful comparison of costs and benefits.

Controlling acid rain would avoid the enormous costs of acidification to Chinese agriculture and forestry projected under the business as usual scenario. Untargeted application of the best available technology across the country would be very expensive ($18 billion a year) and would have a low economic return.[13] Current sulfur dioxide emissions total about 26 million tons a year, and without controls will increase to as much as 45 million tons by 2020. Simulations with the RAINS-Asia model show that an untargeted strategy with low-cost technology would require $6 billion a year to stabilize sulfur dioxide emissions at the 1995 level by 2000 and beyond (map 3). This strategy would avoid damages of some $24 billion in agriculture and forestry. However, the same benefits could be achieved at about half the cost with a targeted application of low-cost controls only to those sources that cause the most damage and only in the regions with the most acid rain (map 4). Such controls include limestone injection for new coal-fired power plants, flue gas desulfurization systems for large industrial boilers, and adoption of low-sulfur fuels for household and transport use.

Although the untargeted strategy would almost eliminate acid rain in China, the costs would be very high. However, at a fraction of the cost the targeted acid rain control strategy described above would produce the same benefits as implied by the government's untargeted emission reduction goal.

Conclusion

The foregoing analysis suggests that there is good news for China's environment. An environmentally sustainable pattern of growth can both increase incomes and improve environmental quality. And with a few crucial adjustments, this future is well within reach of China's current policies and resources. Moreover, the rate of return to these investments is enormous. Each yuan invested in the medium-cost investment scenario will yield 3 yuan in

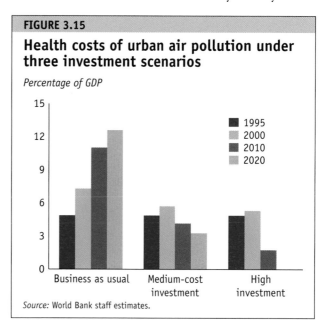

FIGURE 3.14

Particulate concentrations under three investment scenarios

Micrograms per cubic meter

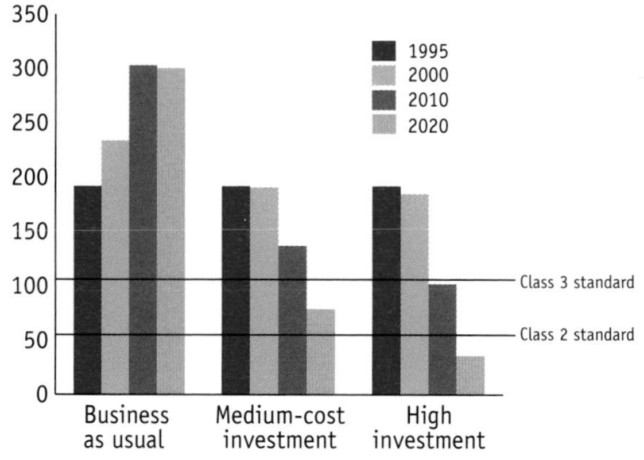

FIGURE 3.15

Health costs of urban air pollution under three investment scenarios

Percentage of GDP

Source: World Bank staff estimates.

Source: World Bank staff estimates.

reduced pollution damages.[14] The returns are even higher if the effects of growing income are considered.

There is a cautionary note in this chapter as well. The business as usual projections indicate that, unless policies are changed, China will not achieve its objectives for air and water quality by 2000. Although its objectives for 2010 might be obtained for certain water pollutants, there are likely to be sharp increases in pollutants before pollution loads return to today's levels. The outcomes of the business as usual scenario are sufficiently worrisome to give pause to environmental policymakers who are tempted to rest on their laurels.

The challenge is to identify policies that will generate the type of growth assumed in the medium-cost and high investment scenarios. The projections raise strategic questions for policymakers:

• How can China reduce its reliance on coal, improve overall efficiency, and adopt cleaner technologies? Needed changes will require more than a gradual shift based on relative costs, preferences for convenience, and changing patterns of fuel use (these effects are already taken into account under the business as usual scenario). It will require nothing less than a determined, systematic set of new policies for developing and using energy.

• What policies can induce enterprises to invest in pollution abatement? As a greater share of output originates in the nonstate sector, creating incentives for nonstate enterprises to invest in protecting the environment is important. This includes incentives to use fewer materials and less energy as well as to control pollution using end-of-pipe measures. Such measures will also reduce the government's investment costs.

• How can China develop environmentally friendly transport modes that reduce pollution and avoid the mistakes made by other East Asian countries? The government's recent decision to phase out lead is an important step that will protect large numbers of Chinese, particularly children, from exposure to lead. In the long run China should implement comprehensive urban transport planning that provides alternatives to private automobiles, traffic jams, and smog.

• Finally, can China develop water policies that protect precious clean water supplies by inducing enterprises and municipalities to invest in abatement efforts as well as conserve?

These issues are the subject of the next four chapters, on energy, industry, urban transport, and water.

Annex 3.1 Modeling industrial output, energy demand, and emissions

The projections made in this report were produced using a model that attempts to capture the impact of economic reform and industrial restructuring on the evolution of industrial output in transition economies. The model, originally constructed for use in Central and Eastern Europe and the former Soviet Union, is not a full general equilibrium model because it is based on exogenous assumptions about macroeconomic performance. Still, it is possible to check the consistency of these assumptions against industry-level projections. The results of this cross-check are used to highlight some of the technical and economic difficulties involved in making medium- and long-term projections for economies undergoing drastic structural changes.

Basic assumptions

The basic assumption driving the model is that over the next two decades transition economies will converge (at varying speeds) toward economic performance and industrial structures similar to those in today's middle-income countries. Differences in initial endowments of capital and natural resources and established patterns of comparative advantage will affect how this process works out. Convergence implies uniformity only to the extent that the legacy of central planning means these countries have similar characteristics that differentiate them from comparable market economies.

The full model is based on an input-output framework with separate matrices of coefficients for old capital and new capital. The coefficients for old capital change gradually over time. The standard input-output convention—that the output of a sector is equal to the demand for goods and services produced by that sector—is followed. As a result the model is (piecewise) linear as far as calculating each year's output is concerned, though the underlying cost functions are not linear. This means that the model can be solved sequentially by matrix techniques. Various adjustment mechanisms have been incorporated to ensure conformity between the macroeconomic assumptions about trade and the sectoral data and assumptions about import and export dependence.

The model has been implemented with the aid of a general matrix programming language, MATILDA (Hughes and Voss 1991), that was primarily designed for modeling based on an input-output framework. The model could easily be translated to a number of other matrix programming languages such as MATLAB or SAS-IML.

Specifications for China projections

As with all linear models, there is a considerable amount of special case detail designed to ensure that the projections do not violate conventional bounds on variables and to prevent various kinds of degeneracy. The China version of the model includes sectoral breakdowns of twenty-four sectors, including twenty manufacturing sectors and four energy sectors (coal, gas, oil products, power and heat), that can be matched with two- to three-digit International Standard Industrial Classification (ISIC) industries (with a few modifications). China's input-output table for 1992 was used as the base year for the model.

The model incorporates assumptions that can be grouped into four major categories: macroeconomic growth (GDP in constant prices, GDP expenditure categories, and adjustment period), power sector development, energy prices, and environmental policies (technological or emission standards, pollution fees). By changing some of the assumptions, the model can test a number of growth scenarios and policy choices, including speed of adjustment, energy prices (for example, fuel taxes), fuel use patterns, emission standards for new technologies, and pollution fees.

Energy coefficients. There are a number of special features concerning the treatment of energy inputs in the model. Over the past thirty years market economies have had an increasing share of electricity in industrial energy despite increases in the price of electricity relative to other fuels. Thus a trend in the share of electricity has been incorporated in the model for all industrial sectors. In the main scenarios this trend includes a higher rate of increase for new capital than for old, reflecting the greater potential for automating Western technologies.

Heat also poses a problem. Institutional arrangements—which are reinforced by statistical conventions in the reporting of data—mean that enterprises purchase (or are recorded as purchasing) heat from electricity utilities and other sources for process use or space heating instead of purchasing primary energy that would otherwise be used to generate heat within the enterprise. This is unlikely to apply to new plants, while old plants will gradually want to control their own use of energy rather than rely on external heat supplies. Thus it has been assumed that heat inputs are zero for new capital and that heat inputs per unit of output for old capital fall by 50 percent by the end of the forecast period.

The model contains a special submodel dealing with the generation of electricity and heat—an important issue because of its effect on primary energy use and emissions. The submodel rests on a simplification of a merit order model for operating electricity plants. First, net electricity demand is adjusted by a factor reflecting transmission losses and own use to give gross electricity demand. The sum of primary electricity production and net imports is deducted to yield the need for electricity from thermal plants. The stock of thermal generating capacity available in any period is used in proportions that reflect the relative advantages of relying on different fuels. New plants have a higher thermal efficiency than old plants and are at the top of the merit order. Old plants are brought on line in accordance with the merit order implied by the assumed utilization shares. The potential output from old capacity depreciates by 3 percent a year, while explicit assumptions are made about investment in new capacity.

Emission factors. There are five matrices representing the emission coefficients for output from old capital and five for output from new capital. For each set, the first matrix represents process emissions per unit of output and thus is applied to total output. The next three matrices represent emissions per unit of fuel used in each sector for three fuels—coal, gas, and oil. Power generation is included as one of the sectors to which these matrices are applied. The final matrix relates to emissions from households and other sources whose volume is linked to population rather than output or inputs. The coefficients represent emissions per 1 million people.

The emission factors for new capital are much lower than those for old capital because modern technology embodies process improvements and better controls that drastically reduce most pollutant emissions. However, various environmental scenarios have been constructed by making alternative assumptions about the stringency of the environmental regulations that must be satisfied by production from new capacity. The base scenario assumes that these regulations reflect U.S.

and EC emission standards that were in force during the early 1980s, which correspond to typical emission factors for U.S. and Western European plants today. This assumption captures the idea that current technologies have been fully adapted to these past emission standards, so that the cost of meeting these standards is small relative to the total cost of new capacity. Other scenarios are constructed by assuming that old capital has to be brought into the emission standards for new capital by the end of the forecast.

Data on actual emission coefficients are difficult to obtain. It is possible to estimate the emission of certain air pollutants linked to energy use, but most of the emission matrices used in the model have been compiled from a wide variety of sources.

Notes

1. The main assumptions about economic growth are taken from the general equilibrium model developed for the analysis in World Bank (1997). Under the assumption that investment will decline from over 40 percent of GDP today to about 34 percent by 2020, consumption will increase from 58 percent of GDP in the mid-1990s to 66 percent in 2020. The share of agriculture in consumption will decline, the shares of industry and construction will remain almost the same, and the share of services will rise. Within the industrial sector the share of food, drink, and tobacco (currently 43 percent) will increase slightly, with offsetting declines in textiles and chemicals.

2. Assumptions in the model about the fuel composition of new generating capacity are based on the analysis in Johnson and others (1996).

3. Energy prices can play an important role in determining both the rate and composition of growth in energy use. In general, Chinese energy prices for industrial users in some parts of the country were close to world market levels during the first half of the 1990s. Since there is no direct link between border prices and domestic wholesale prices, sharp movements in the exchange rate can lead to short-term divergences until domestic prices adjust. Thus in 1994 the wholesale prices of both coal and diesel oil in coastal cities such as Tianjin, Shanghai, and Guangzhou were below the equivalent import parity price at the current exchange rate (after adjusting for calorific value). The high cost of transporting coal in China means that there are large differences between border parity prices for coal in these coastal cities and those in more distant inland cities. Nonetheless, relative prices for steam coal in inland cities were often much lower than would be justified on a simple netback basis using a reasonable (or even high) cost of freight.

4. Demand for gasoline and diesel fuels is quite inelastic with respect to price in the short run—short-run price elasticities typically range from −0.1 to −0.2—but it is much more elastic in the long run (eight to ten years)—long-run price elasticities range from −0.8 to −1.2.

5. This estimate of final energy demand is slightly higher than the baseline scenario developed for Johnson and others (1996), in part because of slower changes in the technical input-output coefficients. For energy unit conversions, see chapter 4, note 1.

6. To allow for the improvements in environmental performance emanating from new capital investment and growth in total factor productivity, the business as usual scenario assumes that average emission coefficients for new investment fall because of new technology. Emissions from the post-1995 component of the capital stock in each sector correspond to those from modern technologies achievable through zero or low-cost environmental improvements. New machinery typically is 25–50 percent more efficient than the existing capital stock and has the kind of emissions performance typical of new investments carried out in industrial countries in the early 1980s.

7. Inevitably, the assumptions underpinning the business as usual scenario do rough justice to the complexities of technical change and environmental progress in different sectors. The most obvious concerns are :

• The scenario makes no allowance for reductions in emission coefficients for the pre-1995 capital stock. In the long run this makes no difference, since such capital will represent a negligible portion of the total capital stock by 2010. In the short term it is not clear whether the assumption of constant emission coefficients over- or understates total emissions because improvements in environmental performance must be set against a general tendency for emissions to increase as plants and equipment get older.

• Presumably, future technical advances will ensure that average emissions from new capital investments made in 2010 that incorporate zero or low-cost controls will be lower than that from new investments made now. Thus the average emission coefficient for new capital should fall over time. Although this assumption is correct, it will be partly or wholly offset by the gradual decline in the emissions performance of the plants and equipment installed in the early part of the post-1995 period. At worst the projections may slightly overstate the level of pollution loads in 2020. These and other uncertainties imply that the projections should be regarded as providing merely a general indication of the trends in total emissions of different pollutants.

8. The sample included all cities with an estimated population of 3 million or more in 1995.

9. This is not to suggest that burning coal in smaller towns and villages has no effect, which is clearly incorrect. The point is that the number of people affected by the smoke generated by burning 1 ton of coal is likely to be significantly less in small towns and villages than in medium-size or large urban areas. Of course, the distinction is blurred where rural areas are very densely populated, as around the periphery of Shanghai, in which case it may be appropriate to treat the whole region as a single conurbation.

10. Indeed, the main effect on urban air quality is likely to be through an increase in the level of fine particulates as a result of the conversion of sulfur dioxide into sulfates in the atmosphere. However, this too is widely dispersed, so the typical impact on a particular urban area will be small relative to the impact of changes in direct emissions of particulates from small sources.

11. According to the National Environmental Protection Agency, China invested about 0.7 percent of GDP in environmental protection between 1991 and 1995, and plans to invest about 1.2 percent of GDP between 1996 and 2000. The investment shares estimated in this report cannot be directly compared with the National Environmental Protection Agency's figures because of differences in coverage. However, it is useful to compare the relative increase in investment in both cases.

12. European Union legislation defines the goal of such policies as the attainment of a "high overall level of protection for the environment." The concept of best available techniques is defined in the European Union's Directive on Integrated Pollution Prevention and Control, formally adopted in September 1996. Note that the concept of best available techniques is not identical to best available control technology because it places much greater emphasis on waste prevention, operational practices, and maintenance, and relies less on end-of-pipe controls. As such, it is much more relevant to industrializing

countries that face severe constraints on capital resources and that need to invest in training as well as cleaner production technologies in developing cost-effective approaches to pollution management.

13. Untargeted emission reduction policies are the same for all sectors and all regions.

14. These estimates are derived by calculating the net present values of annualized costs and benefits of air pollution abatement for the medium-cost and high investment scenarios relative to the business as usual scenario, using a 12 percent discount rate.

References

Hughes, G.A., and S. Voss. 1991. "MATILDA: Reference Manual." University of Edinburgh, Department of Economics.

Johnson, T.M., J.F. Li, Z.X. Jiang, and R.P. Taylor, eds. 1996. *China: Issues and Options in Greenhouse Gas Emissions Control.* World Bank Discussion Paper 330. Washington, D.C.

Marer, P., J. Arvay, J. O'Connor, M. Schrenk, and D. Swanson. 1992. *Historically Planned Economies : A Guide to the Data.* Washington, D.C.: World Bank.

NEPA (National Environmental Protection Agency), SPC (State Planning Commission), and SETC (State Economic and Trade Commission). 1996. *National Environmental Protection Ninth Five-year Plan and 2010 Long-term Goals.* Beijing: China Environmental Science Press.

Steinberg, D. 1992. "Economies of the Former Soviet Union : An Input-Output Approach to the 1987 National Accounts." World Bank, Socioeconomic Data Division, Washington, D.C.

World Bank. 1997. *China 2020: Development Challenges in the New Century.* Washington, D.C.

chapter four

Energy: Coping with Coal

o other major economy is as reliant on coal as China. Current consumption of more than 1.3 billion tons of coal a year is the main cause of China's air pollution and acid deposition. If current trends in energy consumption continue, much of the increase in energy demand over the next twenty-five years will be met by coal. Thus reducing emissions from coal burning is crucial to resolving much of China's air pollution problem. Mitigating the environmental effects of energy consumption will require improving energy efficiency, diversifying energy supplies, and controlling emissions. Energy pricing and pollution regulation will be important since they affect incentives to pursue these goals.

Trends in energy consumption

Two distinct trends in China's energy consumption have developed over the past fifteen years: growth in demand for primary energy has been consistently lower than growth in

GDP (by as much as a factor of two) and final energy demand has been increasingly shifting toward high-grade sources such as electricity and gaseous fuels (natural gas, coal gas, liquefied petroleum gas). Sustaining these trends over the next twenty-five years would go a long way toward reducing China's energy-related pollutant emissions.

Even though China's GDP has quadrupled since 1980, its primary commercial energy demand has only doubled, reaching 1,240 million tons of coal equivalent in 1995, second in the world behind the United States.[1] In addition, rural households burn 250 million tons of coal equivalent of biomass each year, including 173 million tons of dry fuelwood and 298 million tons of dry crop stalks (SETC and CERS 1996).

Increased power plant demand and constrained domestic oil supply raised coal's share in primary commercial energy consumption from 74 percent in 1980 to 78 percent in 1995.[2] This high dependence on coal is unique among the world's major economies (figure 4.1).

Coal is consumed in almost every part of China's economic life. Besides power plants, about 410,000 industrial boilers and 180,000 furnaces and kilns burn coal. Power plants account for about 30 percent of coal consumption, industrial boilers about 30 percent, and furnaces and kilns about 20 percent. Most Chinese urban households use coal stoves for cooking or space heating, often with poor ventilation. And most centralized heating is provided by small coal-fired boilers located in residential or business districts.

The low quality of commercial coal complicates environmental management. Most coal reserves in China have medium ash content and low to medium sulfur content. Because there are few incentives to produce and consume washed coal, most users rely on raw coal. About 20 percent of mined coal is washed, but most washed coal is used in metallurgy. Less than 10 percent of steam coal is washed in China, compared with about 45 percent in the United States and more than 75 percent in Europe. The average ash content of commercial coal has hovered around 20 percent for decades.[3] The high ash content contributes to the relatively low heating value of Chinese steam coal, rated at less than 21 megajoules per kilogram. The average sulfur content of commercial coal is about 1.1 percent, although parts of southwestern China consume coal with a sulfur content as high as 4 percent. By compari-

son, most internationally traded steam coal has 25 percent more heating value per kilogram, 25–50 percent less sulfur, and 50 percent less ash.

High-grade, less-polluting energy sources, such as electricity and gaseous fuels, account for a growing share of final energy demand. During 1990–95 solids (coal and coke) accounted for one-third of the growth in final energy demand, down from two-thirds during 1980–89. The share of solids in final demand fell from 65 percent in 1980 to 60 percent in 1995.

Compared with OECD countries, the sectoral pattern of China's energy demand is highly skewed, with a large industrial share and relatively small transport and commercial shares (figure 4.2). Over time, however, China's sectoral pattern of energy use is expected to become more like that in OECD countries.

Industry accounts for about two-thirds of China's final commercial energy demand. Boilers are the largest industrial energy user, followed by kilns and furnaces. Nonmetal minerals processing (cement, bricks and tiles, and so on), metal smelting and processing (steel, aluminum, copper, and so on), and chemicals manufacturing (fertilizers, acids, soda ash, and so on) use the most energy, accounting for about 60 percent of industrial final coal consumption. Thus emission reduction efforts should focus in these sectors.

Reducing household coal consumption will also be an important part of controlling urban air pollution. Households account for 15 percent of China's final commercial energy demand and for a larger share of

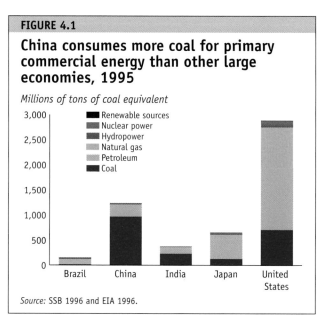

FIGURE 4.1

China consumes more coal for primary commercial energy than other large economies, 1995

Millions of tons of coal equivalent

Legend:
- Renewable sources
- Nuclear power
- Hydropower
- Natural gas
- Petroleum
- Coal

Source: SSB 1996 and EIA 1996.

ambient air pollution. Although increasing gaseous fuel consumption in households has greatly improved indoor and ambient air quality in large urban centers, in 1995 about two-thirds of urban households continued to burn coal for cooking and space heating. More than 100 million rural households still rely on fuelwood or crop stalks for cooking and space heating, but coal use rises with incomes. Commerce, transportation, and agriculture are minor contributors to coal-related air pollution.

Improving energy efficiency

Energy intensity, measured by primary commercial energy consumption per unit of GDP, is a nominal indicator of an economy's energy efficiency. Since 1980 China's energy intensity has fallen by 50 percent, or 4.5 percent a year (figure 4.3). Although a reduction of energy intensity over time is not unusual for mature economies, the consistent and impressive drop in China's energy intensity is unprecedented for an industrializing economy and has been the subject of several studies (World Bank 1993 and 1994; Lin 1996; Sinton 1996).

Efficiency gains in industry are the main reason China's energy intensity has fallen. Slow residential energy growth also has contributed significantly. Structural factors—that is, shifts in final as well as intermediate demand for goods and services, driven by a

changing product mix—are believed to have been the main force behind reduced industrial energy intensity, accounting for about two-thirds of the drop (World Bank 1993). Technical changes—that is, shifts in technologies, processes, and management for producing goods and services—have also been important in lowering industrial energy intensity (World Bank 1994).

Government policies have played a crucial role in promoting these structural and technical changes. Three types of policies have been particularly important:
• *Economic reforms*. Economic reforms since the late 1970s have spurred investment in more efficient production technologies and processes, expanded high-value economic activities, and increased competition. These developments helped make the economy more energy efficient.
• *Energy pricing policies*. Evidence strongly suggests that rising real prices for coal, oil, and electricity have improved resource use and increased conservation among energy consumers (see below).
• *Energy conservation policies*. China has implemented nationwide energy conservation programs since the early 1980s. Although these programs were largely technology-driven, they played a key role in reducing industrial energy intensity during the 1980s (World Bank 1993).

Despite the dramatic fall in energy intensity, China's economy is still among the world's most energy-inten-

---

FIGURE 4.2

Industry uses the most energy in China, 1992

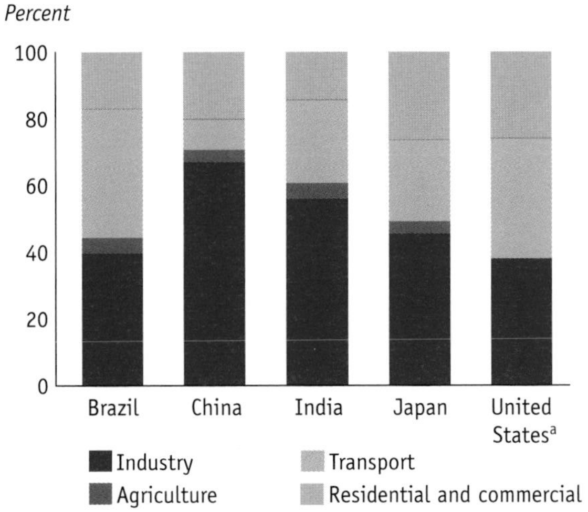

Percent

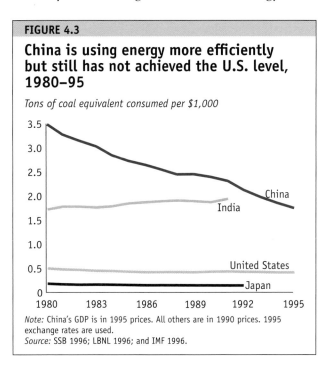

Industry
Agriculture
Transport
Residential and commercial

a. Agricultural energy use is included in industry.
Source: LBNL 1996.

FIGURE 4.3

China is using energy more efficiently but still has not achieved the U.S. level, 1980–95

Tons of coal equivalent consumed per $1,000

Note: China's GDP is in 1995 prices. All others are in 1990 prices. 1995 exchange rates are used.
Source: SSB 1996; LBNL 1996; and IMF 1996.

sive economies. Acknowledging the pitfalls in cross-country GDP comparison, China's energy intensity in 1995 was about four times that of the United States (see figure 4.3). Moreover, industry plays a larger role in China's economy than in Japan and the United States (table 4.1) and still relies on inefficient and small-scale production technologies. China is also much less technically efficient than industrial countries, especially in energy-intensive industries and major energy-consuming equipment (table 4.2). If current trends continue, China's energy intensity will reach 0.586 tons of coal equivalent per $1,000 of GDP by 2020—three times less than in 1995 but still about 40 percent more than the current U.S. level. Thus there is huge potential for reducing China's energy intensity.

China can achieve energy savings through further structural shifts toward high value-added and low energy-intensity products and through industrial modernization. Over time China's economic structure is likely to become similar to that in Japan and the United States, with an increasing role for services (see table 4.1). Services consume far less energy per unit of output.

Technological progress will significantly improve industrial energy efficiency. The potential industrial energy savings identified in table 4.2 could save 250 million tons of coal (at 1995 energy consumption levels), or about 20 percent of current coal consumption—a level that will rise over the next twenty-five years. Investing in these key technologies would generate high returns from saved energy costs and avoided emission control costs.

If the 1986 energy-efficient building standards for heated residential buildings were implemented, new residential buildings with central heating could cut heating coal consumption by more than 30 percent. At a construction speed of 1,456 million square meters a year in 1995, China's building stock will double in about fifteen years. Thus improving the energy efficiency of buildings is crucial to curbing future residential and commercial energy demand.

Diversifying away from coal

As China's economy enters the twenty-first century, the environmental consequences of expanding the energy system based on coal are prompting a search for alternatives. But China is an energy-scarce economy, with per capita energy endowments far below the world average (table 4.3). Although future exploration will increase energy reserves, especially for natural gas, China may have limited scope to reduce its dependence on coal by relying solely on domestic energy resources.

Since 1980 China has increased its reliance on coal to fuel its rapid economic growth. A massive hydroelectric program has been sustained, yet hydropower's overall contribution to the energy supply has fallen. Limited domestic oil resources have been used mainly for transportation and the petrochemical industry, with oil-fired power plants converted to coal-fired during the 1980s and net oil exports ceasing in 1993. Natural gas has played an insignificant role in China's economy, but new exploration may increase its contribution. Nuclear and renewable energy play a small role today but could grow significantly in the future, depending on investment and research and development policies.

China's energy diversification strategy should reflect concerns about energy security and the costs of alternative energy supplies. Continued strong support for oil exploration and development is crucial for meeting future demand. The government also plans to develop domestic natural gas reserves to increase supplies to cities. China will also develop its vast hydropower resources, bringing long-term benefits to hydro-rich southwestern provinces (where large amounts of high-sulfur coal are used now). Efforts should be made to deal with dam-related environmental problems, how-

---

TABLE 4.1
Energy-consuming industries play a large role in China's economy
(percentage of GDP)

Indicator	China, 1995	Japan, 1992	United States, 1993
Total GDP	100	100	100
Agriculture	21	2	2
Industry	48	43	26
Services	31	55	72
Total manufacturing value added	100[a]	100	100
Energy-intensive sectors	35.8	22.9	20.5
Metals smelting and processing	14.2	6.0	4.1
Nonmetal mineral products	9.4	4.3	2.6
Chemical materials and products	9.8	10.0	9.0
Paper products	2.4	2.6	4.8
Rest of manufacturing	64.2	77.1	79.5

a. Based on data for enterprises with independent accounting status.
Source: SSB 1996; World Bank 1995a; and UNIDO 1995.

TABLE 4.2

Adopting best-practice technology could reduce China's coal consumption by 20 percent

Equipment or production process	Share of China's energy consumption	Average efficiency in China	High efficiency in OECD countries	One-time energy savings at 1995 level of energy use
Industrial boilers	30 percent of coal consumption	65 percent	> 80 percent	70 million tons of coal
Coal-fired power plants	30 percent of coal consumption	414 grams of coal equivalent per kilowatt hour	< 350 grams of coal equivalent per kilowatt hour	60 million tons of coal
Steel making	10 percent of total energy use	40 gigajoules per ton of steel	20 gigajoules per ton of steel	60 million tons of coal
Cement kilns	7 percent of coal consumption	170 kilograms of coal equivalent per ton of clinker	100 kilograms of coal equivalent per ton of clinker	30 million tons of coal
Fans and pumps	30 percent of electricity use	75 percent	> 85 percent	30 billion kilowatt hours or 17 million tons of coal
Electric motors	40 percent of electricity use	87 percent[a]	92 percent	18 billion kilowatt hours or 10 million tons of coal
Total savings				247 million tons of coal

a. Average efficiency of mean motor size.
Source: Wang and others 1995; World Bank staff estimates.

TABLE 4.3

China has scarce energy reserves to support its large population, 1994

Country	Raw coal	Crude oil	Natural gas	Hydropower
China				
Total	115 billion tons	3.3 billion tons	1.7 trillion cubic meters	1,923 billion kilowatt hours
Per capita	95 tons	3 tons	1,416 cubic meters	1,603 kilowatt hours a year
United States				
Total	241 billion tons	3.8 billion tons	4.6 trillion cubic meters	376 billion kilowatt hours
Per capita	962 tons	15 tons	18,400 cubic meters	1,504 kilowatt hours a year
World				
Total	1,044 billion tons	141.0 billion tons	142.0 trillion cubic meters	14,546 billion kilowatt hours
Per capita	209 tons	28 tons	28,400 cubic meters	2,909 kilowatt hours a year

Source: LBNL 1996; SETC and CERS 1996.

ever. China's ambitious nuclear energy program must deal with high capital costs, inadequate domestic expertise, a potential lack of indigenous uranium reserves, and the uncertain costs and safety of nuclear waste disposal. China should also invest more to develop its large renewable energy resources. Studies have shown that windfarms, solar home systems, and bagasse cogeneration projects have the potential to become commercially viable on a large scale. Other technologies, such as biogas power and geothermal power, need further research and development before commercial development, but have substantial promise in the long term (World Bank 1996).

International trade is another avenue of diversification, especially in the short to medium term. China has been a net oil importer since 1993 because of rising domestic demand for oil products. Chinese experts estimate that domestic crude oil production will peak at 200 million tons by 2020. But between 1980 and 1995 oil demand increased about 4 percent a year. If these trends continue, total demand for oil will reach 415 million tons by 2020, implying that China will need to import more than 200 million tons of oil.[4] This extrapolated growth in oil demand only partly factors in the potentially huge demand from automobiles and excludes the possibility of using oil as a fuel source for base-load power plants, industrial boilers, and space heating.

China could use energy trade to the advantage of its economy and its environment. In 1995 China spent about $5 billion on 35 million tons of oil and gas imports (SETC and CERS 1996). Net oil and gas imports totaled about 10 million tons, worth roughly $1.4 billion. China's rapidly increasing incomes and export revenues will enable it to import substantially more oil and gas.[5] Given the coastal provinces' wealth and easy access to imports, using imported oil and gas to reduce coal consumption is a realistic and possibly cost-effective option for coastal cities. Establishing a transparent pricing framework for crude oil, oil products, and natural gas is essential to achieving effective and efficient management of the domestic oil and gas market. It is also vital for attracting foreign investment in developing domestic oil and gas resources as well as the infrastructure for imports. If China's oil and gas imports were to rise to 200 million tons by 2020, the cost would be only $28 billion at today's prices, a small portion of its anticipated $700 billion in export earnings in that year.

Controlling emissions: Options and costs

In addition to improving efficiency and developing alternatives to coal, controlling emissions is critical for an energy system that depends on coal. Abatement should focus on industry, power, and residential and commercial sources.

Industry

Industrial boilers and furnaces consume almost half of China's coal and are the largest source of urban air pollution. Low-efficiency cyclones, the most common emissions control equipment for Chinese industrial boilers, remove only 65–85 percent of fly ash. Medium-size and large industrial furnaces often have dust capturing equipment such as bag houses and electrostatic precipitators. But the massive number of small furnaces and kilns, especially those operated by township and village industrial enterprises, generally lack emissions control.

As industry adopts larger and more efficient boilers, such as spreader stokers, dust emissions could triple—or worse. Fluidized bed combustion boilers require stringent dust control, too. Thus the government should strengthen dust removal standards for all boilers to keep dust emissions in check.

Sulfur emissions control for industrial coal combustion would be expensive if only end-of-pipe measures were applied. China has been experimenting with scrubber technology for large industrial boilers and promoting the use of industrial briquettes. Controlling sulfur emissions with briquettes or with lime injection during combustion is relatively cheap and removes 30–50 percent of sulfur. In the long term, fluidized bed combustion boilers, which remove 50–80 percent of sulfur, would be a cost-effective way to reduce sulfur dioxide emissions.

Because China's industrial users normally consume raw coal with an ash content of 25 percent or more, increasing the use of washed coal would substantially lower dust emissions nationwide and sulfur dioxide emissions in regions using high-sulfur coal.

Power

The power sector consumes nearly one-third of China's coal and is a major contributor to acid rain. Many of the thermal plants constructed before 1980 are located near cities and have relatively low smokestacks, and therefore are large contributors to local air pollution. In 1991 only 38 percent of China's power plant boilers were equipped with electrostatic precipitators. Chinese power plants remove an average of 93 percent of the dust from the coal they consume—a low level by international standards.

Few coal-fired power plants in China have flue gas desulfurization equipment. Faced with increasing sulfur dioxide emissions and acid rain in some regions, China has begun implementing various abatement measures in existing and new coal-fired power plants (box 4.1). High-efficiency sulfur scrubbers can increase the initial cost of new power plants by 15 percent or more. In the United States retrofitting old plants for sulfur dioxide control is 20–50 percent more costly than applying similar measures to new plants, implying the benefits of installing sulfur removal equipment at the outset in areas with serious sulfur deposition problems.

The near-term strategy for controlling emissions from the power sector includes:
• Speeding up the adoption of electrostatic precipitators in coal-fired power plants, especially those that are located near cities.

Clean coal technology

Since coal will continue to be a major source of energy over the next twenty-five years, the development of clean coal technologies, especially for power generation and large industry, should be part of China's long-term energy strategy. There are broad applications for coal gasification, fluidized bed combustion, and integrated gasification and combined cycle technologies in industry, power, and the residential and commercial sectors. Technological advances are lowering the costs of these cleaner technologies and making them more competitive in certain areas. Integrated gasification and combined cycle technology has become an increasingly attractive way to generate power in countries where base-load power plants are predominantly coal-fired. In China the capital cost of building a 500 megawatt integrated gasification and combined cycle plant ($850 a kilowatt) is comparable to that of a pulverized coal-fired plant with imported generating units and electrostatic precipitators ($850–950 a kilowatt). Over twenty years the levelized cost of electricity from a 500 megawatt plant using state-of-the-art integrated gasification and combined cycle technology compares favorably with the cost from an advanced pulverized coal-fired plant—and emissions are lower.

The cost competitiveness of integrated gasification and combined cycle power generation improves with plant size and the co-production of chemical products from synthetic gas. Still, the capital cost of a domestic 350 megawatt pulverized coal unit with electrostatic precipitators and flue gas desulfurization ($680 a kilowatt) is much lower than that of an integrated gasification and combined cycle plant with the same output. There is also competition from natural gas–fired combined cycle power generation. Thus a plant's choice of technology will have to be determined on a case by case basis. Over time the wider application of integrated gasification and combined cycle and other clean coal technologies and increasing domestic technology know-how are expected to improve these technologies' competitiveness relative to conventional coal technologies.

Source: Stoll and Todd 1996.

• Focusing sulfur dioxide emissions control efforts on regions suffering from serious acid rain damage, and allowing power plants a choice of abatement measures.
• Tightening emission controls for old plants whose retirement has been postponed (box 4.2).

Residential and commercial sectors

While coal is still the dominant commercial fuel for Chinese households, the increasing use of liquefied petroleum gas, natural gas, and coal gas in urban households has improved air quality in large cities across China. In addition, urban households using coal stoves have benefited from the government's persistent efforts to promote coal briquettes. But burning briquettes is not all that much healthier than using raw coal. Urban household fuel policy should instead accelerate gasification programs.

During the winter much of the population in northern regions still suffers from high exposure to particulate matter, carbon monoxide, and sulfur dioxide because space heating—whether it is household stoves, small boilers, or district heating—is almost entirely dependent on coal. District heating facilities with stringent emission controls are an effective way of controlling space-heating emissions.

Reducing indoor air pollution remains a major challenge for China's rural households. The government's efficient stove program has succeeded in both reducing excess biomass consumption and improving rural indoor air quality (Smith and others 1993). Other cost-effective strategies for rural households include promoting the efficient use of biomass and coal briquettes. Educating rural families about indoor air quality is also important. The provision of low- or zero-emission alternatives such as biogas, liquefied petroleum gas, and small hydropower, solar, and wind energy will largely depend on regional access to these energy sources and should be encouraged by government research and development policies.

Tea boilers and large coal stoves used by cafeterias and restaurants also contribute to ambient air pollution in Chinese cities. A 1994 survey in Beijing found more than twice as many operating tea boilers and work-unit cafeteria stoves as industrial boilers. Switching to coal briquettes and eventually to gaseous fuels is the best way to reduce emissions from these sources.

The special role of energy prices

Prices affect energy intensity and the choice of fuels. Compared with the energy-rich United States, China's energy prices are low (table 4.4). Lower prices may be partly justified by China's relatively low production costs. Over the past decade energy pricing in China has moved from complete controls to a mix of market-driven allocation and government interventions. Govern-

The costs and benefits of emissions control in Shanghai's power sector

Total suspended particulate emissions. About 22 percent of particulate emissions in Shanghai come from the power sector, and more than 90 percent of these emissions come from small power plants (less than 125 megawatts). Retiring these old plants would dramatically cut particulate emissions from the power sector. If retirement of the plants is delayed, complying with emission standards will require installing high-efficiency electrostatic precipitators in these plants.

The cost of removing particulates in Shanghai's seven 25 megawatt units using electrostatic precipitators is just $100–240 per ton of particulates removed (assuming ten years of operation). The cost for 100–125 megawatt units is $300–500 per ton of particulates removed. These costs are much lower than the estimated costs of health and other damages from particulate emissions in Shanghai, which range from $1,200 to $2,600 per ton of particulates.

Using washed coal, especially in small power plants, is also a socially cost-effective option for reducing particulate emissions. Removing particulates using washed coal costs about $720 per ton of particulates removed.

Sulfur dioxide emissions. The power sector accounts for 72 percent of sulfur dioxide emissions in Shanghai. Removing sulfur dioxide emissions costs $100–400 a ton. The lowest-cost options are:

- Installing flue gas desulfurization in power plants that burn coal with a sulfur content of more than 2 percent ($144 per ton of sulfur dioxide removed).
- Using atmospheric fluidized bed combustion technology ($163 per ton of sulfur dioxide removed).
- Using washed coal in existing power plants ($176 per ton of sulfur dioxide removed).
- Using liquefied natural gas–fired combined cycle technology ($100–400 per ton of sulfur dioxide avoided, depending on whether plants are dispatched as base-load or peaking units).

These costs are less than or comparable to the estimated costs of health and other damages of sulfur dioxide in Shanghai—about $390 per ton of sulfur dioxide. Each option could lower sulfur dioxide emissions by 2,000–4,000 tons over 1997–2020. If more sulfur dioxide removal is required, additional and more efficient options will be needed. Applying simplified flue gas desulfurization technology to new coal-fired power plants would cost $225 per ton of sulfur dioxide removed; using wet flue gas desulfurization technology in new plants would cost $310 per ton of sulfur dioxide removed. Pressurized fluidized bed combustion technology costs $210 per ton of sulfur dioxide removed, and integrated gasification and combined cycle technology costs $300 per ton of sulfur dioxide removed.

Source: World Bank staff.

---

ment control varies but is normally associated with the scarcity of specific energy types.

Coal prices are largely decided by the market[6] and vary significantly depending on the destination of the coal.[7] Transportation costs for coal are high in China, accounting for up to 70 percent of the delivery price of coal, compared with 30–50 percent in the United States (X. Wang 1996). Current coal prices are probably close to the direct economic costs of bringing the coal to local markets. The heating value of the coal—and thus, the ash content—is at least partly factored into current prices.

About 70 percent of the domestic crude oil supply is controlled. As a result the average crude oil price is 13–24 percent below international prices. But domestic oil products are priced significantly higher than international spot market prices, giving refineries an unfair profit. As a relatively scarce energy source, domestic crude oil should be priced at least at international levels. The separation of crude oil production from oil refining and the favorable pricing policies for refineries

have created a special interest group within the oil industry that keeps refinery margins high and distribution margins low.

The natural gas supply is fully controlled. The highest-priced gas usually goes to the commercial sector, while the lowest-priced gas goes to fertilizer plants (which consume about one-third of China's natural gas). In 1995 natural gas prices in Sichuan ranged from 470 yuan per 1,000 cubic meters for fertilizer plants to 670 yuan per 1,000 cubic meters for the commercial sector, and were lower in other regions (Q. Wang 1996). Urban natural gas users are probably willing to pay considerably more than prevailing prices. Deregulation of the natural gas industry will speed up the development of domestic natural gas.

In most coastal provinces, where new power capacity has expanded the fastest, electricity prices have been raised to levels approximating marginal costs. Average consumer prices of power in Zhejiang Province at the end of 1994 were about 50 fen per kilowatt hour, compared with the long-run marginal cost of about 45 fen

TABLE 4.4
China's energy prices are lower than those of the resource-rich United States, 1995
(U.S. dollars)

Energy type	China[a]	United States[b]	Spot market[c]
Steam coal (gigajoule)	1.1–2.0	1.2–1.5	~ 1.5
Crude oil (ton)	103	126	118–135
Gasoline (ton)	322–356	382	158
Diesel (ton)	273–290	348	148
Natural gas (gigajoule)	1.4–2.1	1.8–5.5	
Electricity (thousand kilowatt hours)			
Industrial users	26–72	47	
Residential users	26–60	84	

a. Prices include value-added taxes. Calculated using 1995 average exchange rate ($1 = 8.4 yuan). Coal prices are for cities in nonmining areas, and the range roughly indicates different transport costs. Crude oil price is the weighted average of controlled prices and planned prices. Gasoline prices are #90 RON retail prices in different cities. Natural gas prices are for fertilizer plants (low) and households (high) in Sichuan. Electricity prices are for Jinan, Shandong (low) and Shenzhen, Guangdong (high) in 1994.
b. Coal prices are for power utilities (low) and industry (high) and vary depending on quality. Gasoline and diesel prices are the average retail prices of all grades. Average natural gas prices are for power plants (low) and households (high).
c. Coal price is for Australian exports. Crude oil price is for OPEC crude. Gasoline and diesel prices are Rotterdam product prices.
Source: SETC and CERS 1996; EIA 1996.

per kilowatt hour (World Bank 1995a). In interior provinces, by contrast, a large portion of the electricity is produced by plants that were built before 1980. Because these plants effectively have no capital costs, since they were built with nonrepayable central government funding, it has been difficult to raise power tariffs to marginal cost levels in these areas.

Policies for reducing environmental impacts

To achieve their emissions and energy objectives, policies must create strong incentives for improving energy efficiency, diversifying energy sources, and reducing emissions through economic instruments as well as regulations. Key issues include:

• *Energy efficiency improvement.* To improve energy efficiency, policies should promote industrial modernization (cleaner and more efficient production technologies and processes, economies of scale, and better enterprise management), high-efficiency industrial equipment (boilers, electric motors, fans, and pumps), industrial energy conservation (plant or equipment renovation, energy housekeeping, and demand-side management), and energy-efficient buildings (lighting systems, space conditioning efficiency, and building standards).

• *Cleaner coal and energy diversification.* To make coal consumption cleaner and secure the reduced emissions of alternative energy sources, policies should encourage coal washing (using steam coal for industrial boilers and anthracite for residential and commercial users), promote gaseous fuels for residential and commercial use, and support research and development in low- or zero-emission energy sources and technologies. In the short to medium term, increasing the use of imported gas and oil could reduce the growth in demand for coal and consequently overall emissions.

• *Emissions control.* Emissions would fall if high-efficiency dust precipitators were installed in industrial boilers and furnaces, high-efficiency electrostatic precipitators were installed in power plants and large point sources, and sulfur dioxide emissions were controlled in areas burning high-sulfur coal.

To realize these objectives in the least-cost way, the government should develop market-based policies to shift incentives, pursue institutional reforms, and introduce regulatory changes.

Market-based instruments: Energy pricing and taxation

Arguably the most important policy to encourage efficient energy development is to ensure that energy prices at least cover economic costs and then begin to reflect environmental damages through taxes. The government should quickly lift all remaining price controls on coal so that prices reflect its full market value. It is critically important for China to reform the coal allocation and distribution system, which distorts the market and obscures the pricing signal for coal quality.

The case for levying a pollution tax on coal is strong. About two-thirds of China's coal is consumed in the nonpower sectors, with a massive number of small users. Enforcing emissions control among these users is an impossible task for regulatory agencies. As an initial approach, a tax based on the quality of coal (ash and sulfur content) should be introduced. Such a tax would not produce direct and immediate emission reductions, but over time it would lead to increased production and use of better-quality coal as well as raise the demand for more efficient equipment or alternative fuels. Such taxes could be collected by local fuel companies that distribute coal to small users at local markets and to larger users at the mouth of mines.

Eventually, the government should tax coal in a way that reflects its enormous social costs. If health costs were included, the price of coal would increase by 100 percent in Beijing (box 4.3). A tax to mitigate the social costs of coal use could be phased in over time, perhaps three to five years, so that current investment decisions—in boilers, space heating, and so on—could anticipate future prices and adjust technology accordingly.

China's "new plant, new price" policy has helped raise much-needed capital for the country's power development. The overall grid price increased by about 4.5 percent a year (in real terms) during 1990–94, with coastal provinces registering even higher growth.[8] Electricity pricing is rather chaotic but is moving in the right direction. The pricing authorities do not, however, support the idea of increasing electricity prices to control pollution. Because of the difficulties of getting approval for a higher grid purchase price, Chongqing's Luohuang Power Plant often has opted not to run its imported flue gas desulfurization facility. As the government moves to implement its acid rain control program, power plants in the control region will face similar problems. Expenses for environmental control will have to be factored into electricity prices.

Gaseous fuel prices are generally too low. Town gas is normally subsidized, and its prices often do not cover production costs.[9] The same situation applies to centralized heating. Urban households with access to gas and centralized heating usually can afford the full costs of these services. Subsidies not only drain local government budgets but also impede the expansion of these services through private investment. Subsidies for gaseous fuel supply and centralized heating should be removed and tariffs increased to allow for full cost recovery of such projects, essential for improving urban air quality.

Hydrocarbon prices also need to be adjusted. Despite its limited resource base, China prices petroleum products more like the resource-rich United States than energy-scarce East Asia. Although gasoline prices cover the costs of production and importation, they create incentives to expand vehicle use (see chapter 6).

For large point sources, such as power stations and large industrial boilers and furnaces, emissions can and

---

BOX 4.3

Assessing the health costs of coal use in Beijing

Because market-based pricing of coal does not reflect the danger it poses to human health, coal is by far the cheapest fuel for Chinese consumers. The relative price of coal would increase significantly if it were adjusted to reflect the health costs. Using the methods outlined in chapter 2, a range of possible price increases was calculated reflecting low, moderate, and high valuations of these costs. In Beijing including the health costs of coal use would increase the base price of coal by an average of 100 percent, to 400 yuan a ton ($48) on coal with a base price of 200 yuan a ton (see table).

Because space heating accounts for a disproportionately large share of ground-level air pollution in Beijing, particularly during the winter, the health costs of coal use by space-heating boilers is 163 percent of the base price of coal. Damages from the residential (cooking), industrial, and power sectors are lower.

This calculation indicates that taxes could help internalize the health costs of air pollution in Beijing. While the best policy would be to tax the emissions associated with coal burning, this is not feasible for small users. Moreover, the amount of an environmental tax need not be as high as the full externality costs because coal consumption falls as the price increases, reducing the health damage costs, and because abatement costs (emissions control, energy efficiency, fuel switching) are much lower

than externality costs, so a less-than-full tax is sufficient to achieve abatement targets.

Other policies that reduce ambient concentrations of fine particulates from coal consumption in urban areas would have high social rates of return, including efforts to promote district heating, residential gasification, and taller smokestacks, relocating power and industry away from densely populated areas, and installing pollution control equipment.

The market price of coal is less than half its cost to society

(Health costs of coal use in Beijing as a percentage of current coal price)

Sector	Low[a] estimate	Moderate[b] estimate	High[c] estimate	Mean
Power	49	60	173	94
Residential	37	45	129	70
Space heating	85	104	299	163
Industry	62	76	219	119
Total	52	64	184	100

a. Includes mortality (human capital approach) and acute morbidity costs.
b. Includes mortality (human capital approach) and acute and chronic morbidity costs.
c. Includes mortality (willingness-to-pay approach) and acute and chronic morbidity costs.
Source: World Bank staff estimates.

should be taxed. A tax on emissions would discourage investment in small power plants (less than 100 megawatts), which have high emissions per unit of electricity. Taxing emissions would send the right signals for future investment.

Institutional reform: Making the right investment decisions

One of the main pitfalls of China's investment policy is that it inadvertently encourages investment in small-scale and often inefficient technologies. Examples of such investment pervade energy-intensive industries such as cement, steel, fertilizers, and power generation.[10] This misguided investment orientation is as much a capital constraint problem as it is an institutional problem. The provision that any project costing more than 30 million yuan has to be approved by the State Planning Commission overburdens the central government and discourages investors from building large-scale plants. Local and regional protectionism also undermine the economic efficiency of investment. Reforming state lending institutions and the financial market and giving local governments more autonomy in making investment decisions would facilitate investment in modern technologies and increase the economies of scale of new plants.

The government should strengthen its financial support to research and development for low- or zero-emission energy alternatives, especially in promising renewable energy sources and clean coal technologies. These new technologies would improve and preserve China's environmental quality in the long run and put China in a favorable position to deal with global environmental issues.

Regulations and standards: Overcoming market failures

Left to themselves, markets will not necessarily make environmentally sound choices. Regulations and standards, whether market-based or command-and-control, must be enforced to ensure that the economy achieves its environmental goals. Air pollution control regulation and enforcement must be tightened to improve urban air quality. A high priority for regulation is to introduce standards to control fine particulate emissions from space heaters, industrial boilers, kilns and furnaces, and power plants. To promote cleaner

and more efficient coal use, the levy for dust emissions should be adjusted to reflect the cost of abatement necessary to meet emission standards.

China should implement its plans for sulfur dioxide and acid control regions as proposed in the revised 1995 National Air Pollution Prevention and Control Law. The government should introduce market-based regulations to reduce sulfur dioxide emissions from coal-fired power plants and large industrial sources, especially in areas that have high ambient concentrations and high sulfur deposition. Regulation should allow emission sources to choose among least-cost abatement measures.

Energy efficiency standards for heated residential buildings are needed to curb long-term space-heating coal demand and should be implemented without further delay. Introducing energy efficiency standards for major household appliances is also important for China's consumer market.

China is at an energy and environmental crossroads. The performance of its huge and growing energy system will have a tremendous effect on air quality. If it can harness the market, strengthen institutions, and regulate more effectively, China can ensure that more efficient and environmentally friendly technologies are adopted. The potential benefits for the future quality of life are enormous.

Notes

1. The conversion factors are 1 ton coal equivalent = 29.310 gigajoules; 1 ton Chinese average raw coal = 20.934 gigajoules; 1 ton Chinese average crude oil = 41.868 gigajoules; 1,000 cubic meters Chinese average natural gas = 38.98 gigajoules; and 1 kilowatt-hour = 3.6 megajoules.

2. The share of coal cited here is higher than the official figure because a different conversion factor for electricity was used.

3. According to a user survey, the ash content of steam coal is usually more than 25 percent (NEPA 1996).

4. In 1995 the world oil trade (imports and exports) was 1,800 million tons. U.S. net imports were 390 million tons, and Japan's were 280 million tons (BP 1996).

5. Importing 100 million tons of crude oil would have cost about $14 billion in 1995. Total imports that year were $132 billion (SSB 1996).

6. Steam coal used by power plants is priced 20 yuan a ton lower than the market price because electricity prices are kept relatively low by the government (Q. Wang 1996).

7. For example, in June 1994 the mine-mouth price of Datong mixed coal was 128 yuan a ton. The same coal retailed for 230 yuan a ton in Shanghai, 280 yuan a ton in Guangzhou, and 340 yuan a ton in Xiamen (LBNL 1996).

8. Assessed based on the national average of power company sales prices adjusted by an implicit GDP deflator.

9. In 1995 town gas (from coal gasification) prices ranged from 0.3–0.7 yuan a cubic meter in twelve of fifteen Chinese cities, while the costs of gasification alone ranged from 0.6–1.0 yuan a cubic meter. Only three cities charged more than 1.0 yuan a cubic meter for town gas (Q. Wang 1996).

10. About 20 percent of newly installed thermal power capacity in China between 1991 and 1995 was in 6–75 megawatt units (Jianglong 1997). During this period China's cement production increased by 235 million tons, more than 70 percent of which came from plants with average capacity of less than 50,000 tons a year, one-tenth the average U.S. plant size (Sinton 1996).

References

BP (British Petroleum). 1996. *BP Statistical Review of World Energy.* London.

EIA (Energy Information Agency). 1996. United States Energy Data, International Energy Data. EIA Web site: http:/www.eia.doe.gov. U.S. Department of Energy, Washington, D.C.

IMF (International Monetary Fund). 1996. *International Financial Statistics Yearbook.* Washington, D.C.

Jianglong, Zhou. 1997. Personal communication. Beijing Economic Research Institute of Water Resources.

Johnson, T.M., J.F. Li, Z.X. Jiang, and R.P. Taylor, eds. 1996. *China: Issues and Options in Greenhouse Gas Emissions Control.* World Bank Discussion Paper 330. Washington, D.C.

LBNL (Lawrence Berkeley National Laboratory). 1996. *China Energy Data Book.* Berkeley, Calif.

Lin, Xiannuan. 1996. *China's Energy Strategy: Economic Structure, Technological Choices, and Energy Consumption,* Westport, Conn.: Praeger Publishers.

Liu, Feng. 1996. "Energy Use and Conservation in China's Residential and Commercial Sector." Report LBL-33867. Lawrence Berkeley National Laboratory, Berkeley, Calif.

MOEP (Ministry of Electric Power). 1995. "China's Electric Power Industry 1995." Beijing.

NEPA (National Environmental Protection Agency). 1996. *Handbook of Industrial Pollutants Emission Factors.* Beijing.

SETC (State Economic and Trade Commission) and CERS (China Energy Research Society). 1996. *China Energy 1996.* Beijing.

Sinton, Jonathan. 1996. "Energy Efficiency in Chinese Industry: Positive and Negative Influences of Economic System Reforms," PhD dissertation. University of California at Berkeley.

Smith, Kirk, Shuhua Gu, Kun Huang, and Daxiong Qiu. 1993. "One Hundred Million Improved Cookstoves in China: How Was It Done?" *World Development* 21(6).

SSB (State Statistical Bureau). 1996. *China Statistical Yearbook 1996.* Beijing: China Statistics Press.

Stoll, Harry, and Douglas M. Todd. 1996. "Current IGCC Market Competitiveness." General Electric Company, Schenectady, New York.

UNIDO (United Nations Industrial Development Organization). 1995. *International Yearbook of Industrial Statistics 1995.* New York.

Wang, Qingyi. 1988. *Zhong Guo Neng Yuan* (Energy in China). Beijing: Metallurgical Industry Press.

———. 1996. Personal communication. Ministry of Coal, Beijing.

Wang, Qingyi, Jonathan Sinton, and Mark Levine. 1995. "China's Energy Conservation Policies and Their Implementation, 1980 to the Present, and Beyond." Lawrence Berkeley National Laboratory, Berkeley, Calif.

Wang, Xiaodong. 1996. "China's Coal Industry: Moving toward Market Economy." World Bank, Summer Internship Project, Washington, D.C.

World Bank. 1993. *China: Energy Conservation Study.* Report 10813-CHA. Washington, D.C.

———. 1994. "Energy Efficiency in China Technical and Sectoral Analysis." *Issues and Options in Greenhouse Gas Emissions Control.* Washington, D.C.

———. 1995a. *Investment Strategies for China's Coal and Electricity Delivery System.* Report 12687-CHA. Washington, D.C.

———. 1995b. *World Tables.* Washington, D.C.

———. 1996. *China: Renewable Energy for Electric Power.* Report 15592-CHA. Washington, D.C.

MAP 1
Sulfur deposition in China, 1995

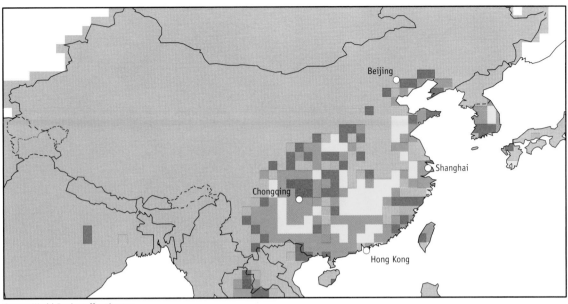

Source: World Bank staff estimates.

MAP 2
Sulfur deposition in China under the business as usual scenario, 2020

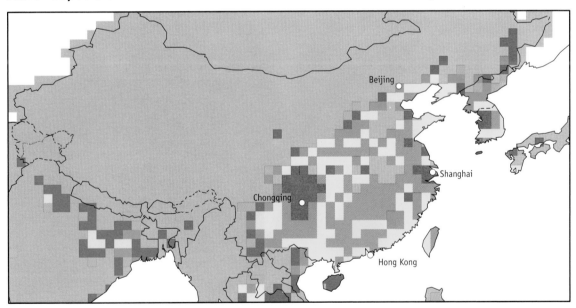

Source: World Bank staff estimates.

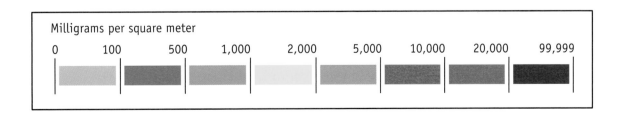

Milligrams per square meter

| 0 | 100 | 500 | 1,000 | 2,000 | 5,000 | 10,000 | 20,000 | 99,999 |

MAP 3

Sulfur deposition in China with untargeted control efforts, 2020

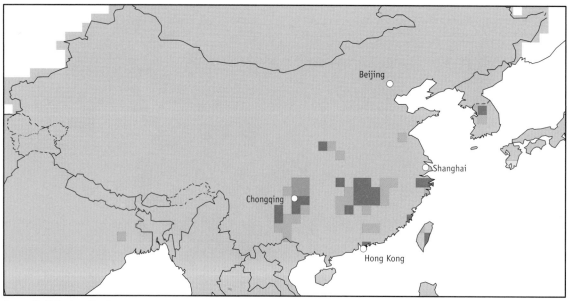

Source: World Bank staff estimates.

MAP 4

Sulfur deposition in China with targeted control efforts, 2020

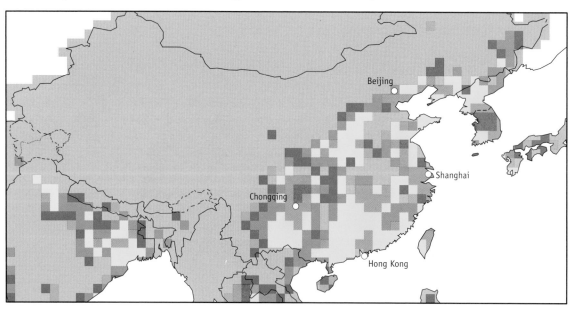

Source: World Bank staff estimates.

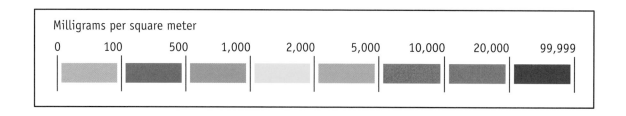

Milligrams per square meter

0   100   500   1,000   2,000   5,000   10,000   20,000   99,999

MAP 5

Water quality in China's major water systems, 1981

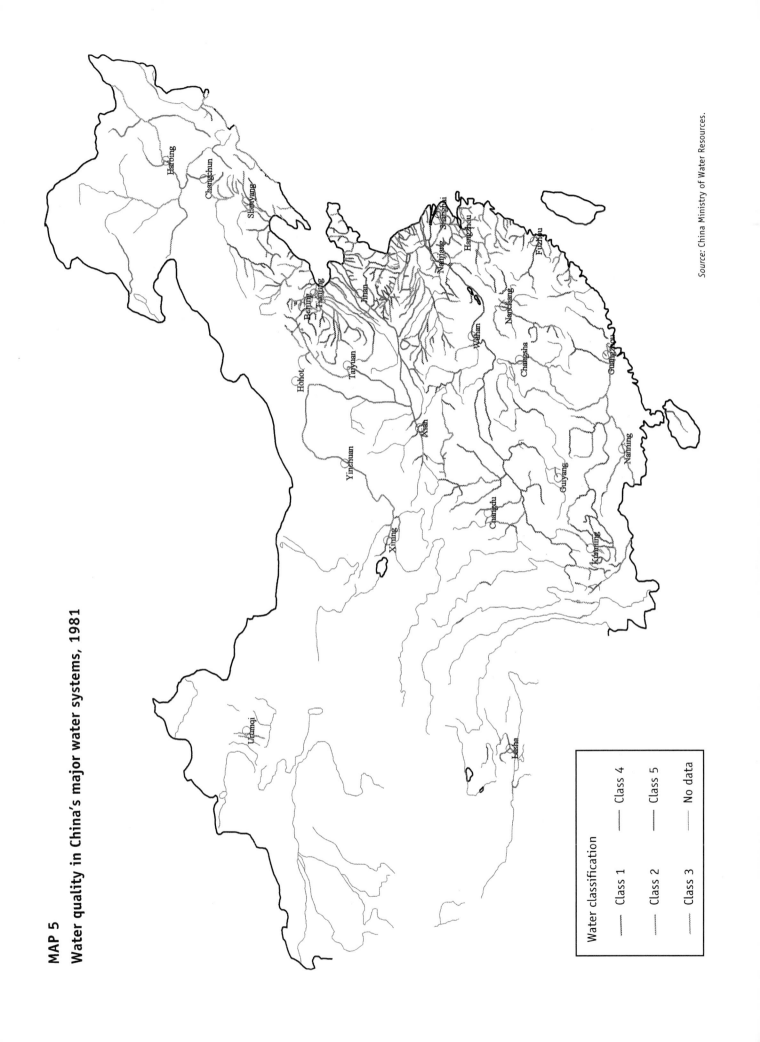

Water classification

—— Class 1
—— Class 2
—— Class 3
—— Class 4
—— Class 5
—— No data

Source: China Ministry of Water Resources.

MAP 6

Water quality in China's major water systems, 1995

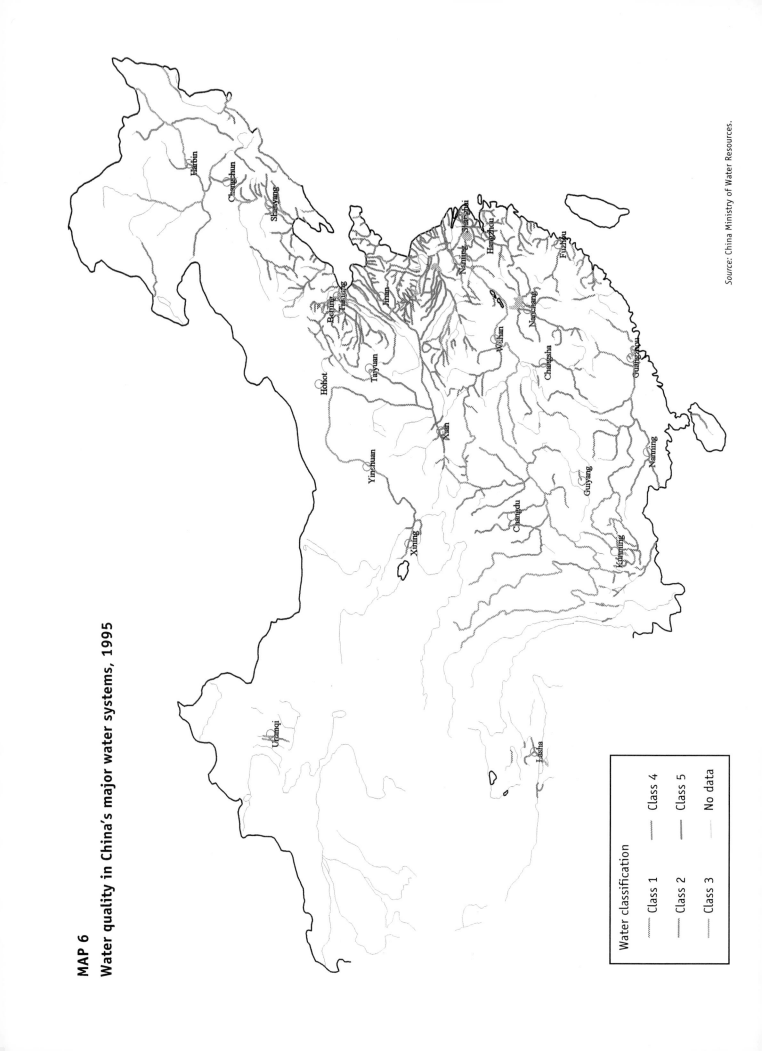

Water classification

Class 1

Class 2

Class 3

Class 4

Class 5

No data

Source: China Ministry of Water Resources.

chapter five

Industry: Creating Incentives to Abate

Industry is the engine of growth for the Chinese economy, accounting for 48 percent of GDP in 1995. In the 1990s the output of China's 10 million industrial enterprises has increased by more than 15 percent a year.

But this engine of growth is highly polluting. The National Environmental Protection Agency estimates that discharges from industry—including power plants—account for more than 70 percent of national pollutant emissions. Many polluting industries are located in densely populated metropolitan areas, exposing urban residents to serious health risks.

The prospect of continued rapid industrial development creates the opportunity for cleaner industrial technologies to replace today's highly polluting ones. With growth expected to average nearly 7 percent a year over the next twenty-five years, 80 percent of China's industrial production capacity remains to be built. To realize the potential for cleaner production, China must revamp its regulatory system to harness market forces and reorient new investment to safeguard the environment.

Pollution control policies

Over the past decade the Chinese government has enacted more than a dozen laws affecting the environment, and these now provide a comprehensive legal framework for China's environmental management. The pollution control system encompasses a wide range of environmental regulations and standards and a nationwide enforcement network of local environmental protection bureaus working with sector agencies.

China's pollution control policies are based on three main principles administered through eight regulatory programs.[1] The three principles emphasize:
- Prevention first, then prevention with control.
- Polluters pay.
- A strong regulatory framework.

Pollution prevention focuses on new pollution sources, which are regulated through environmental impact assessments and the Three Synchronizations program.[2] Pollution from existing sources is regulated through pollution levies, pollution discharge permits, and mandatory pollution controls. These five programs have played a major role in containing emissions from regulated industrial enterprises.

Prevention and control efforts have targeted mainly state-owned enterprises. However, rapidly growing township and village industrial enterprises have emerged as an increasingly important source of pollution (table 5.1). These enterprises' pollution is generally unregulated.

Prevention

Environmental impact assessments and the Three Synchronizations program are the main mechanisms for ensuring that new industrial facilities invest in pollution control equipment. The program mandates that new industrial enterprises and existing enterprises that plan to expand or change their production process receive environmental impact assessment approval for their plans from the National Environmental Protection Agency or local environmental protection bureaus. In this way pollution control facilities are integrated with enterprises' development plans.

The effectiveness of both mechanisms is limited because they focus on source-specific emission standards. As a result, even if all new projects in a region meet the standards, ambient air quality can deteriorate because no limit is placed on total pollutant discharges. Another weakness is their strong bias toward end-of-pipe treatment. Both the assessments and the Three Synchronizations program focus on making projects meet emission standards through onsite hardware installation rather than by pursuing cost-effective strategies at the regional level. As a result broader, more flexible approaches—such as resource pricing and incentive-based control—are precluded. The National Environmental Protection Agency has proposed changing both mechanisms to facilitate its efforts to control the total quantity of pollutant emissions.

Control

Pollution levies, discharge permits, and mandatory pollution controls help contain pollution from existing sources. Pollution levies were introduced to create an economic incentive for industrial enterprises to comply with emission and effluent standards, to raise revenue for investment in industrial pollution control, and to provide financial support for regulatory activities.

The levies combine two elements: noncompliance fees assessed on pollutant discharges that exceed emission and effluent standards, and fines and other charges assessed on violations of the levy regulations. Of the fees collected, 80 percent are used for grants and low-interest loans to pollution control projects and 20 percent are retained by local environmental protection bureaus to support administrative and monitoring activities.

The pollution levy system has several design deficiencies:
- Pollution fees, established at the national level, are often lower than the marginal cost of abatement required to meet emission standards. As a result many industrial enterprises choose to remain noncompliant and pay the levy on their excess pollution. In addition, since the early 1990s inflation has lowered the real value of the levies because they are not indexed, and local governments are not allowed to make their own adjustments.
- Levies are based only on the pollutant that exceeds its standard by the greatest amount, rather than on all the pollutants that exceed their standards.

TABLE 5.1

Township and village industrial enterprises are becoming a major source of industrial pollution, 1989 and 1994

(tons)

Type of enterprise	Chemical oxygen demand		Smoke dust		Processing dust		Sulfur dioxide	
	1989	1994	1989	1994	1989	1994	1989	1994
Regulated enterprises	6.8	6.8	8.5	8.1	7.6	5.8	10.5	13.4
Township and village industrial enterprises	1.8	5.3	5.4	8.0	4.7	5.8	3.6	5.5

Source: NEPA 1995 and 1996.

- Levies are assessed only on above-standard emissions, so they provided no incentive for firms to abate emissions below the standards.
- Levies are based on pollutant concentrations rather than on total discharges to the environment.

Despite these problems, pollution levies, combined with subsidized loans from the levy fund and other subsidy programs, have created incentives to control pollution.

Discharge permits were introduced in 1987 to compensate for the lack of mechanisms to control total pollution loads. Two basic schemes have emerged from trial implementation: capacity-based permits allocate pollution loads to industrial enterprises based on the assimilative capacities of the local atmosphere and receiving waters, and goal-based permits allocate pollution loads to industrial enterprises based on environmental goals not related to assimilative capacities.

The government has experimented widely with permits to control water pollution but has not used them much to control air pollution. The main problems with the water pollution discharge permits are:

- Conflicts between fixed quantity (mass-based) discharge permits and programs based on concentration standards, such as the environmental impact assessments, the Three Synchronizations program, and the pollution levies.
- The inadequacy of monitoring and administrative capacity for implementing the discharge permit system. As a result discharge permits have not realized their potential for controlling total pollution loads and inducing cost-effective abatement.

Mandatory pollution controls under threat of closure have been used as a last resort to force highly polluting enterprises to adopt control measures. Many of these enterprises are old plants using obsolete technologies, so such forced investment for end-of-pipe control has often wasted scarce capital by inducing pollution abatement at excessively high cost.

Monitoring and enforcement

National and local environmental regulations and standards are enforced primarily by municipal and county environmental protection bureaus that report to local governments. The bureaus' ability to monitor compliance is seriously constrained by limited financial and human resources. In general, medium-size and large enterprises are monitored by the bureaus once or twice a year. Small enterprises are monitored less often, and continuous monitoring of wastewater flows and smokestack emissions occurs rarely, if ever.

Although China's environmental standards are consistent with international standards, compliance with these standards is greatly undermined by inadequate enforcement. Weak monitoring, excessive local government intervention (to protect enterprises), and insufficient public involvement contribute to lenient enforcement. According to recent surveys, during 1993–94 only 36–49 percent of industries were in compliance with their environmental impact assessment and the Three Synchronizations program. The compliance rate for mandatory pollution controls is 80 percent. Among enterprises or work units that are required to pay pollution levies, 60 percent pay in full and on time. In general, compliance is higher among larger enterprises (Millison 1997).

Determinants of industrial pollution

The severity of pollution in different parts of China is largely determined by variations in the pollution intensity—pollution per unit of output—of industrial enterprises. To identify potential mechanisms for reducing pollution, this section analyzes the determinants of pollution intensity using data from regulated firms. Focusing on variations in five provinces—Liaoning, Beijing, Shanghai, Guangdong, and Sichuan—we trace the effects of reforms and structural change since 1987, particularly trade reform.

Aside from variation among different types of industries, China's provinces have different levels of pollution because the pollution intensity of plants in the same industry varies considerably. Since 1988, however, pollution intensity has fallen substantially in all five provinces (figure 5.1). Particularly sharp reductions have occurred in Guangdong and Sichuan.

Great variation persists across provinces, however. Holding the sectoral share of industry constant, factories are dirtiest in Sichuan and cleanest in Shanghai. Three basic differences across provinces explain such variation: degree of regulation, pressure applied by neighboring communities, and industrial characteristics.

Regulation

Enforcement of pollution standards varies greatly across China's provinces. For example, actual collections of pollution levy assessments (or *effective levies*) per unit of wastewater discharge above the legal standard vary greatly because of variations in enforcement (figure 5.2). Effective levies per unit of air pollutant discharge above the standard also cover quite a range (figure 5.3).

Using the effective levy as a proxy for all the regulations applied to industrial enterprises, we studied the relationship between regulatory strictness and pollution intensity. Controlling for other factors, effective levies have had significant effects on the pollution intensity of production across China's provinces. Each 1 percent increase in the effective water pollution levy yields an 0.8 percent decrease in chemical oxygen demand intensity per unit of output. It does not matter whether the increase in the levy originates from increased penalties or tightened enforcement. Each 1 percent increase in the effective air pollution levy yields an 0.3 percent decrease in sulfur dioxide pollution intensity, an 0.8 percent decrease in particulate pollution intensity from coal burning, and an 0.4 percent drop in particulate

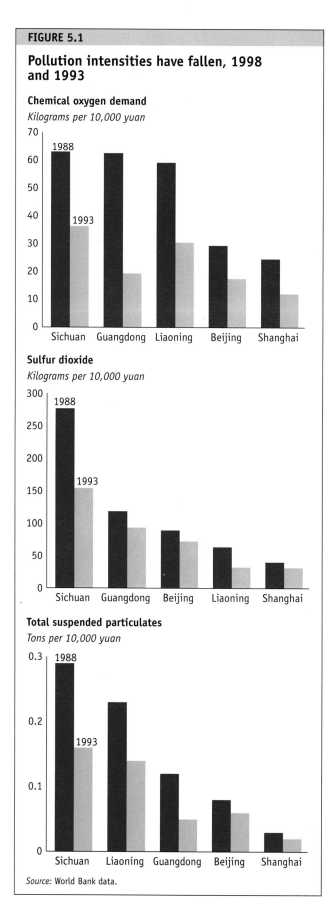

FIGURE 5.1

Pollution intensities have fallen, 1998 and 1993

Chemical oxygen demand
Kilograms per 10,000 yuan

Sulfur dioxide
Kilograms per 10,000 yuan

Total suspended particulates
Tons per 10,000 yuan

Source: World Bank data.

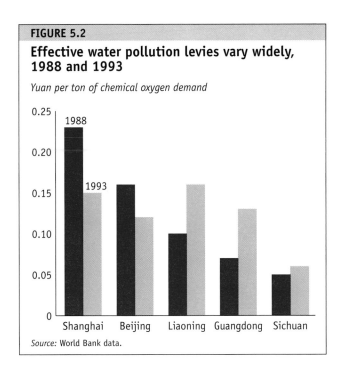

FIGURE 5.2

Effective water pollution levies vary widely, 1988 and 1993

Yuan per ton of chemical oxygen demand

Source: World Bank data.

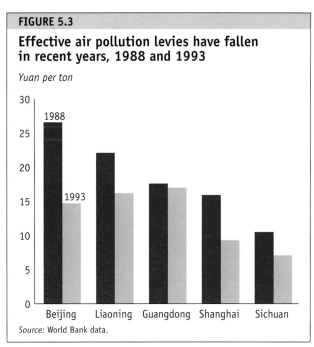

FIGURE 5.3

Effective air pollution levies have fallen in recent years, 1988 and 1993

Yuan per ton

Source: World Bank data.

pollution intensity from production processes. Pollution intensities are significantly lower in provinces with higher effective levies, and since 1987 they have fallen in provinces where levies are rising. Thus regulation makes a significant difference in lowering pollution intensities.

China's provincial regulators have pursued very different regulatory policies. Beijing, for example, has the highest effective levies and among the lowest pollution intensities; the converse is true for Sichuan. Notably, however, Shanghai does not conform to the general pattern: its effective air pollution levy seems entirely out of line, both with its effective water pollution levy and its status as the area with the lowest pollution intensity.

Community influence

Whether pollution charges or fines are enforced partly depends on the extent of the damage and the community's capacity to influence the authorities. As local pollution loads increase and the exposed population grows, the damage to human health increases. These factors, especially in wealthy provinces, trigger increased enforcement. Educated citizens are also more likely to complain about pollution.

It is no accident that Shanghai, a highly developed province, registers the most citizen complaints (figure

5.4). Areas with high illiteracy are notably silent when it comes to complaining about pollution; uneducated citizens are less likely to be aware of environmental problems and less willing to confront the authorities. As complaints rise, so do effective levies for both air and water pollution. As in other countries, citizen feedback is a powerful independent force for environmental improvement.

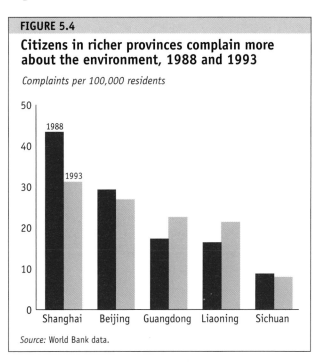

FIGURE 5.4

Citizens in richer provinces complain more about the environment, 1988 and 1993

Complaints per 100,000 residents

Source: World Bank data.

Plant and industry characteristics

Three enterprise-related factors are also important determinants of a province's pollution intensity. First, some industrial processes generate large volumes of harmful pollutants, so the composition of industries affects pollution. And because some sectors face higher abatement costs than others, cost-sensitive enterprises are more resistant to pollution control measures. Second, because end-of-pipe pollution abatement has significant economies of scale, large plants have lower unit abatement costs and respond more readily to regulatory incentives. They also tend to be more technically efficient, since their size allows them to spread overhead costs for skilled personnel across many units of production. Finally, ownership makes a difference. China's state-owned enterprises generate a lot more pollution than their nonstate counterparts. One reason is their lower operating efficiency, which affects both abatement costs and the generation of waste residuals in production (Dasgupta and others 1996). In addition, state enterprise equipment and technologies tend to be older. And state enterprises are less sensitive to price signals, including pollution charges or fines, because they often operate at the pleasure of government authorities and have privileged access to finance.

When the effects of these three factors are analyzed together, for both water pollution (chemical oxygen demand) and air pollution (sulfur dioxide) the share of production in large plants is the most significant factor in determining overall pollution intensity (figure 5.5). The share of production in state-owned plants also carries a lot of weight. Among industries the main contributors to higher intensity vary by pollutant.

Growth and the environment: Reforms and structural change since 1987

China's economic reforms over the past decade have increased the role of price signals in the economy, eased constraints on enterprise growth, and reduced barriers to competition from new domestic (often nonstate) firms and foreign trade. These developments have had three positive effects on the environment. First, average plant scales have increased: the share of large plants in industrial production grew from 30 percent in 1988 to 37 percent in 1994, with a particularly rapid increase in

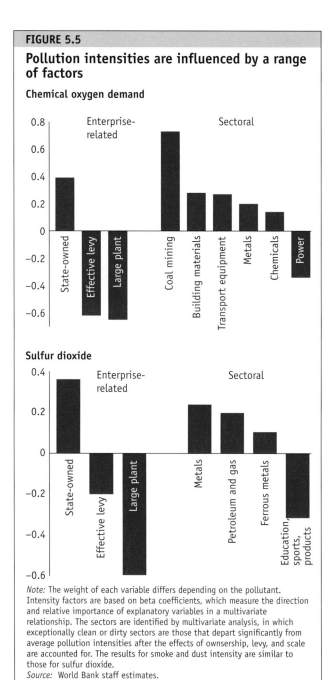

FIGURE 5.5

Pollution intensities are influenced by a range of factors

Chemical oxygen demand

Sulfur dioxide

Note: The weight of each variable differs depending on the pollutant. Intensity factors are based on beta coefficients, which measure the direction and relative importance of explanatory variables in a multivariate relationship. The sectors are identified by multivariate analysis, in which exceptionally clean or dirty sectors are those that depart significantly from average pollution intensities after the effects of ownership, levy, and scale are accounted for. The results for smoke and dust intensity are similar to those for sulfur dioxide.
Source: World Bank staff estimates.

Shanghai (from 41 percent to 58 percent). Since larger plants pollute less per unit of output, increasing scale has brought environmental benefits. Second, the role of state enterprises has diminished, again with a particularly rapid change in Shanghai (from a share of state enterprises in total enterprises of 70 percent in 1988 to 54 percent in 1994). Since nonstate firms tend to pollute less per unit of output, the decline in state enterprises has been accompanied by a significant drop in pollution intensity. Third, rapid growth has brought structural change in industry. The share of particulate-intensive industry has declined significantly, while the shares of chemical oxygen demand– and sulfur dioxide–intensive industry have remained about constant. Reforms have also had a notably clean effect in the five provinces covered earlier. The share of pollution-intensive sectors has dropped in all cases for water pollution and in all but one case (sulfur dioxide in Liaoning) for air pollution. The largest changes have occurred in Beijing for chemical oxygen demand–intensive sectors and in Guangdong and Shanghai for sulfur dioxide– and particulate-intensive sectors.

Trade reform also has had a positive effect on China's environment. Increased production efficiency in export-oriented enterprises and more rapid absorption of newer, cleaner technologies from OECD sources have offset any tendency for China to specialize in pollution-intensive industries because it has weaker environmental regulation and lower-cost heavy raw materials than many of its trading partners.[3]

Moreover, there is no evidence of any long-run shift toward a pollution-intensive industry structure. The shares of the five most polluting industries—chemicals, pulp and paper, nonferrous metals, ferrous metals, and nonmetallic minerals (principally cement)—in total industry have declined (figure 5.6; see Hettige and others 1995 and Mani and Wheeler 1997). Another useful indicator is the trend in net imports (imports minus exports) of products from these sectors. Trade liberalization would accompany a shift toward net exports of these products if any specialization in pollution-intensive industries were emerging. In fact, real net imports of these products have increased considerably since 1987 (figure 5.7).[4]

Overall, trade liberalization has been accompanied by expanded output of goods that are less pollution-intensive. Of course, all industries experienced rapid growth over the past ten years, including pollution-intensive ones. But cleaner industries have grown faster, both in domestic production and in China's international trade.

Market-based regulation during this period failed to reinforce the positive effects of economic reforms, however. Despite damaging air pollution since 1987, real effective air pollution levies have fallen in the five provinces covered earlier (see figure 5.3). The decline is particularly striking in Beijing and Shanghai. Real effective water pollution levies have also fallen in both provinces, while they have risen sharply in Liaoning and Guangdong (and more modestly in Sichuan). Although they have affected plant-level pollution intensities, different levies have not influenced location decisions: there apparently has been no regulation-induced migration of dirty sectors within China (Lucas 1996).

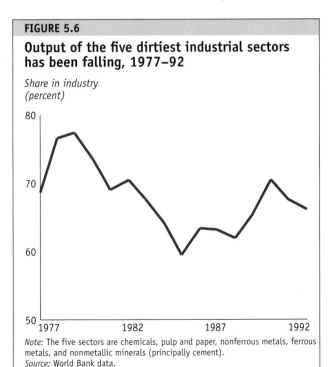

FIGURE 5.6

Output of the five dirtiest industrial sectors has been falling, 1977–92

Share in industry (percent)

Note: The five sectors are chemicals, pulp and paper, nonferrous metals, ferrous metals, and nonmetallic minerals (principally cement).
Source: World Bank data.

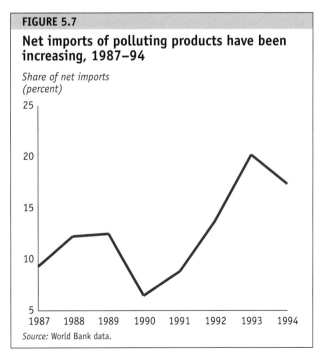

FIGURE 5.7

Net imports of polluting products have been increasing, 1987–94

Share of net imports (percent)

Source: World Bank data.

Since 1987 declining real air and water pollution levies in Beijing and Shanghai have created incentives for increased pollution intensity. Thus the observed declines in intensity should be attributed to countervailing shifts in scale, ownership, and sectoral composition induced by economic reforms as well as by other regulatory instruments. For air pollution intensity the same conclusion follows for Liaoning, Guangdong, and Sichuan. For water pollution, on the other hand, there has been a sharp drop in intensity in these three provinces because higher levies and economic reforms have been mutually reinforcing.

Future pollution scenarios

Assuming that economic reforms will continue, recent regulatory and industrial trends can be used to project their future effects. While several policies can increase pollution abatement, we have used the effective pollution levy to illustrate the potential impact of policy reforms under three scenarios for 1997–2020 for five important cities—Beijing, Shanghai, Shenyang, Guangzhou, and Chongqing. The three scenarios are:

• *Economic reform and the regulatory status quo.* Economic reforms have accounted for much of the decline in air and water pollution intensities since 1987. Even if regulations were not tightened, growth and economic reform would still drive changes in sectoral composition, scale, and ownership. But will reform be sufficient to achieve China's objective of holding industrial pollution to 1995 levels?

• *Tighter regulation: "Polluters pay some."* Tightening regulations would create incentives to reduce pollution intensity and would complement the effects of economic reform. Under this scenario air and water pollution levies would increase by 5 percent a year. Such growth is equivalent to the increase in Sichuan's effective water pollution levy over the past five years and is less than the increases in Guangdong and Liaoning. By 2020 the real levy would be four times today's level.

• *Strict regulation: "Polluters pay more."* Under a third scenario levies would increase by 10 percent a year. By 2020 levies would be thirteen times their current level.

Economic reform and the regulatory status quo would probably hold the chemical oxygen demand load from regulated enterprises to a modest increase (figure 5.8). This is remarkable because the industrial base will have

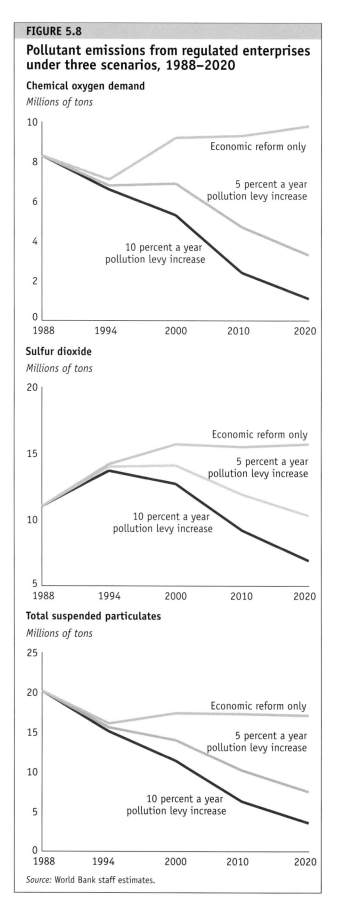

FIGURE 5.8

Pollutant emissions from regulated enterprises under three scenarios, 1988–2020

Source: World Bank staff estimates.

nearly quintupled by 2020. For sulfur dioxide, in the reform-only scenario, continued decline in the air pollution intensity of industrial production is not enough to offset emission increases caused by industrial growth.[5]

Increasing the pollution levy by 5 percent a year—that is, merely continuing recent trends in water pollution regulation—has a dramatic effect on chemical oxygen demand discharges. Projected emissions would fall by more than half, from about 7.0 million tons in 1995 to 3.3 million tons in 2020. A 5 percent annual increase in the air pollution levy would cause sulfur dioxide emissions to fall by a third and particulate emissions to fall by half. Were the levy to increase by 10 percent a year, sulfur dioxide and particulate emissions would each fall by three-quarters.

All three scenarios suggest that the solutions to China's industrial pollution problems are within reach. If chemical oxygen demand regulations were tightened substantially, organic water pollution would fall sharply in regulated industries. Better regulation would also lower sulfur dioxide emissions, significantly improving air quality in all five cities. A 10 percent increase in the pollution levy would prevent further deterioration from the 1993 level in Shenyang and Beijing, while Shanghai, Guangzhou, and Chongqing would experience strong improvements in air quality. Particulate loads behave similarly. For the 5 percent and 10 percent levy scenarios, the projected trends generally replicate the sharp drop in particulate loads projected at the national level. The exception is Shenyang, which experiences only a modest decline. Experiences elsewhere indicate that such charges can be an effective way of achieving pollution reduction targets (box 5.1).

Is stricter regulation worthwhile?

Reducing pollution loads in China would save lives, improve the health of China's people, and increase the productivity of workers (see chapter 2). In the five cities studied here, increasing the air pollution levy by 10 percent a year rather than relying solely on economic reforms would save the lives of more than 4,000 people each year. Compared with the status quo scenario, projected deaths from sulfur dioxide pollution in 2020

BOX 5.1

Global experience with pollution charges

Many countries have experimented with pollution charges. The performance of these charges leads to several conclusions about their effective use:

- For controlling emissions from a large number of small sources in areas where command-and-control regulations are expensive to enforce, fuel taxes and water charges have an advantage over emission and effluent charges. Fuel taxes have a particular advantage as a proxy for the air pollution charge. Unlike emission charges, fuel taxes are easy to collect and can be used to control small sources such as vehicles and small boilers. When air pollution is closely associated with the pattern and type of energy use, fuel taxes also create an incentive to improve energy efficiency and use better-quality fuels.
- Effluent charges based on water use, including user charges for wastewater treatment, can be effectively applied to all categories of consumers in areas where water consumption is metered. However, possible equity effects on large low-income households should be assessed and mitigated.
- Emission charges can be effectively applied to a limited number of pollutants and to large polluters because the expected gains in pollution reduction justify the cost of monitoring and collection.

- Emission charges are most effective when implemented as part of broader pollution control programs with clear ambient air quality targets for a specific area or watershed and emission reduction targets for particular sources. Successful implementation of such programs requires adequate institutional support (for example, establishing river basin agencies or urban air quality councils) and strong political commitment.
- Emission charges should provide an incentive for the targeted level of pollution abatement in an area or watershed. Low emission charges that are introduced primarily to raise funds do not have a comparative advantage over other charges and taxes, given their higher administrative complexity and costs.
- User charges for municipal or collective wastewater treatment should be applied wherever such treatment takes place and should recover the cost of the service to the largest extent possible. It is important to define the size of collective plants and the level of treatment adopted when optimizing the wastewater treatment strategy across all significant sources in a watershed.
- If revenue from pollution charges is earmarked for environmental funds, it is important to have a coherent, transparent, and accountable allocation system with clear financing objectives and priorities.

Source: World Bank staff.

TABLE 5.2

Projected annual deaths from industrial sulfur dioxide pollution under three scenarios

City	Economic reform only	5 percent a year pollution levy increase	10 percent a year pollution levy increase
Beijing	4,500	3,200	2,400
Chongqing	3,200	2,600	2,200
Shenyang	1,200	900	600
Shanghai	1,100	700	700
Guangzhou	300	200	100

Source: World Bank staff estimates.

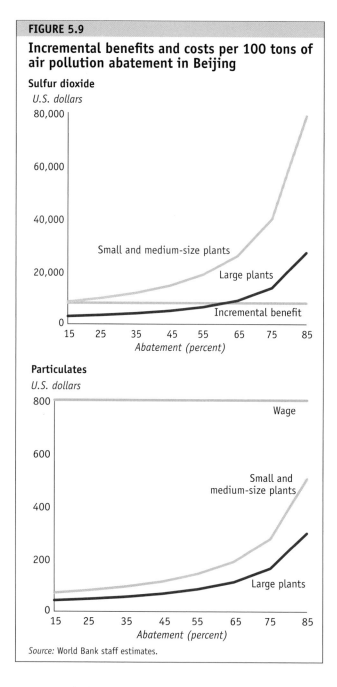

FIGURE 5.9

Incremental benefits and costs per 100 tons of air pollution abatement in Beijing

Sulfur dioxide

Source: World Bank staff estimates.

would be 67 percent lower in Guangzhou, 50 percent lower in Shenyang, 47 percent lower in Beijing, 36 percent lower in Shanghai, and 31 percent lower in Chongqing (table 5.2).

Although saving lives by reducing air pollution is a laudable objective, expenditures on pollution abatement must be judged against other expenditures that might also improve the quality of life. China remains a poor country in which hundreds of millions of citizens have many unmet basic needs. Public investment in health facilities and education can also yield major health benefits, and direct investment in productive capital can improve health by increasing incomes. Thus any analysis of air pollution abatement must consider its costs as well as its benefits.

The analysis has to be undertaken at the margin. The cost of abating China's industrial pollution rises steadily at the margin with the degree of abatement. At some level of abatement further pollution reduction may no longer be warranted because the same resources could be used for other investments with greater life-saving potential. Using Beijing to illustrate our approach, we address this issue with a systematic assessment of abatement benefits and costs.

Because marginal abatement costs vary by pollutant, sector, plant size, ownership, and degree of abatement, it makes little sense to talk about abatement costs in a general way. The question for Chinese regulators is: which facilities should be regulated, and how much abatement is socially desirable? To reflect the overlap between particulates and sulfur dioxide as sources of fine particulates, abatement costs for both pollutants are shown in figure 5.9.[6] Large plants are a major source of air pollution in Beijing, so the marginal abatement costs for these facilities

are particularly interesting. The marginal abatement cost curve for 100 tons of sulfur dioxide abatement by large plants in Beijing ranges from $2,900 at 15 percent abatement to $27,000 at 85 percent.[7] The marginal abatement cost curve for large plants crosses the $8,000 incremental benefit line at around 60 percent abatement; by implication, the marginal abatement cost curve for nonstate large plants would cross the incremental benefit line at a much higher abatement level. The marginal abatement cost schedule for smaller facilities crosses the incremental benefit line at around 15 percent abatement.[8]

For particulates the numbers are much more attractive. Marginal abatement costs scarcely rise above $500 for each 100 tons abated, even at 85 percent abatement for smaller plants. Costs are even lower for large plants. Marginal abatement costs for our other four cities differ somewhat but have the same order of magnitude.

A useful contrast is provided by the National Environmental Protection Agency's current estimates of actual abatement levels in the five cities (table 5.3). According to the agency, Beijing's current abatement rate for industrial sulfur dioxide is 2.5 percent. To derive a conservative benefit-cost assessment, we adopted the lowest rate of abatement—15 percent—that is included in figure 5.9. At 15 percent abatement, the incremental cost of abating 100 tons of sulfur dioxide pollution is estimated to be $2,860. Taking this to be the incremental cost of saving a life, we use our estimated incremental lifesaving benefit ($8,000) to calculate a social rate of return to abatement of about 180 percent.

It appears that the proposed sulfur dioxide levy of $24 a ton will induce less than 15 percent abatement in large plants.[9] To achieve a 60 percent abatement in large plants, the levy has to be raised to $80 a ton in Beijing.

The estimated rate of return to abatement, while clearly attractive, is extremely conservative in terms of the lifesaving value of reducing air pollution in Beijing. For a large nonstate plant our econometric results imply a marginal abatement cost of $332 for each 100 ton reduction in sulfur dioxide emissions when the abatement rate is 10 percent. If—as seems likely, given overall abatement rate of 2.5 percent—large nonstate plants in Beijing are abating less than 10 percent of sulfur dioxide emissions, then the failure to enforce greater abatement is equivalent to valuing a Beijing citizen's life at less than $500 at the margin. Thus continued inaction means ignoring a public investment with a social rate of return in excess of 1,500 percent. The opportunities for such high returns may be quite rare in China's urban health sector.

Toward effective and efficient industrial pollution control

China has made considerable efforts to solve its industrial pollution problems, and these efforts have slowed the growth of major pollutants. But even though it has

TABLE 5.3
Industrial sulfur dioxide abatement in five cities, 1993
(tons)

City	Sulfur dioxide produced	Sulfur dioxide emitted	Abatement (percent)
Shenyang	196,100	144,700	26.2
Shanghai	393,700	356,700	9.4
Chongqing	544,800	494,800	9.2
Guangzhou	165,100	151,500	8.2
Beijing	208,900	203,700	2.5

Source: NEPA 1994.

become more comprehensive, the regulatory framework still lacks enforcement capacity and design coherence. The transitional state of the economy requires China's environmental management agencies to deal with the lingering problems of the centrally planned system as well as the new challenges of the emerging market regime.

The National Environmental Protection Agency has set ambitious targets for controlling China's industrial pollution.[10] Realizing those targets will require China's pollution control authorities to harness the market, strengthen regulations, and reform institutions so that new investment will be channeled to cleaner production technologies and better pollution control.

Harnessing the market and strengthening regulations

The increasing market orientation of the industrial sector offers an opportunity to use market-based pollution controls more effectively. But changes and reforms in the existing regulatory system, if not properly introduced, could actually have an adverse effect on China's environmental management. Thus it may be necessary to tighten existing controls at the same time market-based instruments are being introduced. Recognizing that effective pollution control depends on local enforcement, the following recommendations are made with respect to national policies.

Redesign environmental impact assessments and the Three Synchronizations program. New approaches are required, both to incorporate environmental impact assessments and the Three Synchronizations program into China's overall pollution control strategy and to increase their capacity to make the industrial sector adopt cleaner production technologies. Both programs should be revised to reflect the assimilative capacities of local and regional airsheds and watersheds, rather than continuing the current focus on source-specific discharge standards. Integrating both programs with a market-based discharge permit system would encourage enterprises to adopt least-cost abatement measures.

Reform the levy system. Achieving pollution control objectives will require increasing pollution charges. The National Environmental Protection Agency has proposed a tenfold increase in the air pollution levy; such an increase would go a long way toward reducing air pollutant emissions (see figure 5.8). Higher levies are needed both to lower current emissions and to finance the large investment required to achieve desired ambient air quality in Chinese cities (box 5.2).

Changing the design of the levy system would increase its effectiveness. Because pollution levies are assessed only on above-standard discharges, emissions are "free" for polluters until the standards are reached. Moreover, effluent charges are based on the pollutant that exceeds the standard by the greatest amount and do not reflect the relative risk of individual pollutants. For both air and water discharges it would be sensible to consider adopting a complete charge system that would target specific pollutants and be assessed on all types of pollution. A volume-based levy system would also be easier to integrate with the discharge permit system.

Reforming the management of the levy funds is also crucial to realizing the levy system's potential. Returning

---

BOX 5.2

The optimal air pollution levy in Zhengzhou—and its implications for China

Zhengzhou is the capital of Henan Province, in China's central region. With a 1993 population of 1.8 million people and an average industrial wage of 3,350 yuan a year, Zhengzhou is representative of China's large cities. Its industry emits about 45,000 tons of sulfur dioxide every year. It is estimated that more than 400 people a year are dying from sulfur dioxide–related pollution in Zhengzhou, and thousands are suffering from serious respiratory illness.

Environmental regulators could reduce pollution and improve health in Zhengzhou by increasing the air pollution levy. But how far should they go? The optimal levy should be set at the point where the incremental benefits of abatement are equal to the incremental cost of achieving it. Our estimates of the current situation in Zhengzhou, and the implications of raising the levy to reduce sulfur dioxide emissions, are shown in the figure.

For this analysis we have assumed a statistical life value of $8,000 a person. At the city's current atmospheric sulfur dioxide concentration, abating 1 ton of sulfur dioxide saves about 0.63 percent of a statistical life, yielding a benefit of about $50. By contrast, the incremental cost of abating 1 ton of sulfur dioxide at current emission levels is estimated to be about $1.70. Thus there is huge social return to abating more pollution. As the figure shows, additional abatement remains socially profitable at the margin until 73 percent of current emissions are eliminated. The levy that would induce this reduction is about $90 a ton, at the intersection of the marginal benefit and cost curves. This is the optimal levy for Zhengzhou, since a lower levy would leave socially profitable abatement opportunities unexploited and a higher levy would impose an abatement cost higher than the social gain from further pollution reduction.

Sulfur dioxide abatement yields enormous social returns

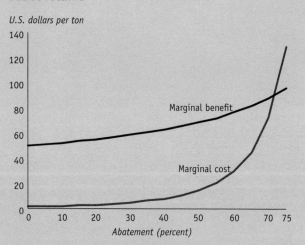

U.S. dollars per ton

The proposed levy of $24 a ton for sulfur dioxide emissions would achieve about 55 percent abatement in Zhengzhou. Current inaction is equivalent to valuing the life of an average urban resident at about $270 ($1.7/0.0063) at the margin. This figure seems ludicrously low relative to the loss of a life, with the associated pain and suffering and the elimination of a lifetime's contribution to China's economic output. Even our suggested life value of $8,000 is a conservative number that only takes account of forgone economic output. But to meet even this conservative standard, our result suggests that the proposed sulfur dioxide levy should be increased to about $50 per ton of sulfur dioxide emissions in Zhengzhou—and, by implication, in the rest of urban China.

Source: World Bank staff.

part of the levy to enterprises (in the form of low-interest loans for pollution control) reduces the political difficulty of levy enforcement. But financial oversight must ensure that the money is well spent. At a minimum, the funds should be pooled and allocated competitively based on the highest marginal abatement per unit of expenditure. In addition, the funds should not be used to support local environmental protection bureaus, which should be funded from general government revenues.

Fast-increasing pollution loads from township and village industrial enterprises deserve special attention. The recent closing of thousands of heavily polluting township and village industrial enterprises probably could have been avoided if past regulation had covered these enterprises. In any case, the long-term solution to regulating these myriad, price-sensitive enterprises lies

in market-based approaches. The government should start experimenting with a regulatory system based on full-cost pricing of natural resources and environmental taxes, while continuing to strengthen existing regulatory measures.

Continued economic reform is needed to preserve past environmental gains. The impact of economic reforms—and accompanying technical and structural change—on the air and water pollution intensity of Chinese industry has been powerful. Reform-induced changes in sectoral composition, ownership, and scale of production have compensated for much of the increase in total output during the past decade. Continued reforms can have similar mitigating effects over the next few decades. Rapid industrial growth that is not accompanied by further changes in ownership and production

---

BOX 5.3

ISO 14000—The new international standard for environmental performance

Certification by the International Standards Organization (ISO) is essential for manufacturing and service firms seeking rapid growth in the global marketplace. Many leading international firms will only do business with subcontractors that have satisfied the ISO's universally recognized performance standards. In the past ISO benchmarks were applied to overall management and quality control. The new ISO 14000 standard, however, explicitly responds to environmental concerns by defining requirements for an environmental management system.

Because the standard is new, some components are still being negotiated. But the core elements of the certification process are now in place. To be certified under ISO 14000, firms must:

• Undergo an initial review by management to identify environmental issues of concern (for example, excessive use of polluting inputs, potential for a serious environmental accident).

• Establish priorities for action, taking into account factors such as local environmental regulations and potential costs.

• Develop an environmental policy statement, to be signed by the chief executive officer.

• Set performance targets based on the policy statement (for example, reducing heavy metal emissions by 50 percent by some date).

• Implement the environmental management system as part of reaching the performance targets.

• Measure performance, review implementation, and make appropriate adjustments to ensure that targets are met.

These "verifiable core elements" of the environmental management system are referred to as ISO 14001. To achieve them,

firms can follow a series of guidelines that provide operational instructions. General auditing principles for tracking performance have already been published as ISO 14010, ISO 14011, and ISO 14012. More specific measures will be detailed in ISO 14030, on environmental performance evaluation, which was reviewed by the ISO Technical Committee in April 1997. Other components of the ISO 14000 environmental certification process are still being developed. These include life-cycle assessment, environmental labeling, and environmental aspects of product standards. All standards developed under ISO 14000 will be reviewed every five years for possible modification.

Although it is new, ISO 14000 is already affecting the environmental stance of firms in both industrial and developing countries. In Europe businesses are rapidly adopting an eco-management and audit regulation that incorporates the ISO 14000 principles and requires firms to publicly report their environmental performance. To remain competitive in Europe, Japanese companies are aggressively pursuing ISO 14000 certification. Most U.S. firms will undoubtedly follow.

Among developing countries, Mexico's Ministry of the Environment has already incorporated ISO 14000 into its regulatory system. The ministry is offering Mexican firms major incentives to adopt ISO 14000, including accelerated depreciation on environmental equipment (for tax purposes) and eligibility for simplified licensing procedures. And both Indonesia and the Philippines plan to incorporate ISO 14000 standards into their new programs for public disclosure of firms' environmental performance.

Source: World Bank staff.

scale, by contrast, would produce far greater pollution loads than those contemplated in this report.

An incentive-based discharge permit system would reduce control costs while achieving local or regional pollution goals. China should move quickly to adopt its overall pollution control strategy by implementing an incentive-based discharge permit program. The integration of such a program with existing pollution control programs will be a major task for China's regulatory agencies.

Enforcement should be targeted at low-cost sources. The cost analysis in this chapter shows why targeting is a good idea: large polluters are easier to monitor and have far lower unit abatement costs. Thus the key to cost-effective pollution control in China's cities is targeted enforcement of higher abatement standards for large facilities.

Adequate compliance monitoring is essential to achieving China's pollution control targets. Compliance monitoring should be strengthened by increasing financial support for equipment and skilled personnel. Random monitoring should increase and more enterprises should be monitored in order to raise the pressure on polluters to abate. Continuous monitoring of large point sources should become mandatory.

Reforming institutions

Institutional obstacles have impeded the implementation of China's environmental programs. These obstacles have been particularly serious for programs requiring coordination between government agencies and between governments that share natural resources (such as water) or face regional pollution problems (such as acid rain). Limited public participation also undermines environmental management.

Institutional responsibility for pollution enforcement should be more clearly assigned. Local governments need to have more control over managing local and regional environmental quality. However, the National Environmental Protection Agency should continue to serve as the central agency for supervision and monitoring. This two-tier arrangement will prevent local complacency about pollution control and ensure effective management of cross-regional pollution issues.

As its importance in global trade grows, China will be subjected to increasingly stringent international standards, including environmental performance and production standards such as ISO 14000 (box 5.3). In response, China should establish an institutional framework for incorporating ISO 14000 into its clean production strategy, starting with trial certifications in the short term and expanding to cover major industries in the medium term.

Local communities should have ways to voice their environmental concerns. In the long run community voice may be the most important way of controlling pollution. The strength of regulation in China's provinces is largely determined by each province's social and economic development. Poor communities with low education levels are far less able to promote their environmental interests than their more developed counterparts. Ultimately, China's environmental interests will be best served by rapid economic development, coupled with concerted efforts to inform and empower poor communities. Nongovernmental organizations will also play an important role in future pollution control efforts.

Notes

1. The eight programs form the core of China's environmental protection activities: environmental impact assessments, the Three Synchronizations program, pollution levies, pollution discharge permits, mandatory pollution controls, centralization of pollution control, the goal-responsibility system of environmental protection, and the quantitative appraisal system of urban environment comprehensive control.

2. The Three Synchronizations program mandates that a project and its pollution control facility be designed, built, and operated in synchronization.

3. Martin, Huq, and Wheeler (1993) find that more open developing economies absorb clean technologies in metals and paper production much more rapidly than their less open counterparts.

4. Finally, we have used the available data to calculate China's consumption-production ratio for the output of the five dirtiest sectors. If China's comparative advantage lies in "dirty" production, then two things should be true. First, this ratio should be less than 1, since China should be a net exporter of pollution-intensive goods. Second, the ratio should fall over time because production for export should increase faster than production for domestic consumption. The evidence is mixed but generally is not consistent with the hypothesis that China's comparative advantage is in pollution-intensive production. Although China's consumption-production ratio has fallen since 1987, it has remained well above 1 throughout the period. This supports the story in figure 5.6: China has been and remains a net importer of pollution-intensive goods.

5. We used projected pollution levies and shares of large plants,

state enterprises, and "dirty" and "clean" sectors to project national and provincial pollution intensities. We assumed that provincial intensities are the same as the pollution intensities of our five cities (not an unreasonable assumption since these cities are the main industrial centers of their provinces and in two cases—Beijing and Shanghai—are identical to their province).

To estimate city emissions, we multiplied projected provincial industrial output by the city's share of provincial industrial output in 1993. Thus we assumed that the growth of industry in the province's main industrial city is identical to the growth of industry in the province. To obtain projected particulate and sulfur dioxide loads for China and the five cities, we multiplied projected pollution intensities by projected industrial outputs. For sulfur dioxide the available information allowed us to go several steps further. We calculated sulfur dioxide concentrations in each city's atmosphere by adjusting the previous year's concentration at 0.51 times the rate of change in the estimated pollution load. This adjustment reflects the results of an econometric analysis of the relationship between emissions and atmospheric concentrations in a large number of Chinese cities. Next we estimated the changes in mortality rates that are induced by changes in atmospheric sulfur dioxide concentrations. For this exercise we relied on recent empirical studies in Beijing and Shenyang by X. Xu and others (1994). Finally, we multiplied estimated concentrations by projected city populations to project deaths from air pollution.

6. Marginal abatement costs for Beijing are estimated by combining the results of an econometric abatement cost study of Chinese industry with current sectoral and ownership data for that city.

7. This calculation averages the distribution of production between state and nonstate enterprise plants in Beijing. Because nonstate plants are less polluting than state plants, a targeted regulatory strategy could exploit this difference to capture the largest returns from stricter enforcement for the least-cost plants.

8. How should lifesaving be valued for comparison with abatement costs? A useful benchmark is provided by the average wage of a worker in Beijing, which was about $800 (6,526 yuan) in 1995. A baseline estimate of the loss to society when one worker dies from respiratory disease is the presented discounted value of the annual wage over a working lifetime. For Beijing this is about $8,000 in present value terms at a 10 percent discount rate. We should stress that it is a very partial index of loss, for two main reasons. First, it takes no account of pain and suffering. Secondly, it focuses exclusively on mortality,

although considerable losses are also associated with working days lost to nonfatal respiratory disease.

9. The National Environmental Protection Agency has been conducting trial implementation of a sulfur dioxide levy in selected cities. The 200 yuan per ton of sulfur dioxide levy is equivalent to $24 per ton of sulfur dioxide (1995 exchange rate).

10. The targets are contained in the State Environmental Protection Ninth Five-Year Plan and 2010 Long-Term Goals.

References

Dasgupta, Susmita, Mainul Huq, David Wheeler, and C.H. Zhang. 1996. "Water Pollution Abatement by Chinese Industry: Cost Estimates and Policy Implications." Policy Research Working Paper 1630. World Bank, Washington, D.C.

Hettige, Hemamala, Paul Martin, Manjula Singh, and David Wheeler. 1995. "The Industrial Pollution Projection System." Policy Research Working Paper 1431. World Bank, Washington, D.C.

Lucas, Robert. 1996. "Environmental Regulation and the Location of Polluting Industry in China." World Bank, Policy Research Department, Washington, D.C.

Mani, Muthukumara, and David Wheeler. 1997. "In Search of Pollution Havens? Dirty Industry in the World Economy, 1960–95." World Bank, Policy Research Department, Washington, D.C.

Martin, Paul, Mainul Huq, and David Wheeler. 1993. "Process Change, Economic Policy, and Industrial Pollution: Cross-Country Evidence from the Wood Pulp and Steel Industries." Paper presented at the annual meetings of the American Economic Association, Anaheim, Calif., January.

Millison, Dan. 1997. Personal communication. Ecology and Environment, Inc., Lancaster, N.Y.

NEPA (National Environmental Protection Agency). 1994. *China Environmental Yearbook 1994.* Beijing.

———. 1995. *Environmental Pollution Protection Policies for TVIEs.* Beijing: China Environmental Science Press.

———. 1996. *China Environmental Yearbook 1996.* Beijing.

Xu, Xiping, J. Gao, D. Dockery, and Y. Chen. 1994. "Air Pollution and Daily Mortality in Residential Areas of Beijing, China." *Archives of Environmental Health* 49(4): 216–22.

chapter six

Urban Transportation: Controlling Cars

Although China's motor vehicle fleet is relatively small, these vehicles are a significant source of urban pollution. Ambient hydrocarbon, carbon monoxide, and ozone concentrations in major cities exceed Chinese standards and are approaching international levels. Mobile sources—especially automobiles—account for a large share of these emissions. Although the number of motor vehicles in Beijing is only about one-tenth of that in Los Angeles and Tokyo, vehicle emissions are about the same for all three cities (Walsh 1996).

The automobile fleet, though small relative to both income and population, is growing rapidly. Since the late 1970s vehicle registrations have grown by 12–14 percent a year. If incomes continue to rise, urban densities fall, and public transit worsens, urban automobile ownership could reach 85–130 vehicles per 1,000 people by 2010, and possibly twice this range by 2020.

This pattern of growth would make traffic jams and smog a way of life. In recent years the number of automobiles has grown nearly twice as fast as the growth in road construction, leading to increasing congestion and much lower average driving speeds. Lower driving speeds increase emissions per kilometer because motors are less efficient at lower speeds and because they take longer to travel the same distance.

If rapid urbanization and motorization are not carefully managed, these trends could produce environmental problems that would rival those of Bangkok, Jakarta, Mexico City, and São Paulo. These cities are shackled with high health costs associated with exorbitant pollution loads, with reduced labor productivity because workers are caught in traffic jams, and with insatiable demands for increased public investment in roads (Faiz 1990).

China can avoid this future. Its limited dependence on automobiles, high urban densities, and extensive administrative capacity will enable it to shape its urban environmental destiny. Doing so will require using both the conventional pollution control policies of other countries—emissions controls and the like—and the full range of policies that shape the future of cities.

Where is China now?

Motor vehicles are a major source of emissions in large Chinese cites (table 6.1), and without new policies their contribution will increase. In Beijing, for example, mobile sources accounted for 40–75 percent of carbon monoxide, hydrocarbon, and nitrogen oxide emissions in 1989. In Guangzhou traffic-related emissions accounted for 87 percent of carbon monoxide and 67 percent of nitrogen dioxide concentrations in 1993.

The national motor vehicle fleet grew from 3 million in 1985 to more than 10 million by the end of 1994. Still, in 1994 there were only 8 motor vehicles per 1,000 people. Private automobile ownership is quite low—less than 2 vehicles per 1,000 people—but has been growing by almost 50 percent a year.

Despite rapid growth, China's motorization rate is low for its level of income relative to other East Asian economies (figure 6.1). Two factors make China different. Higher residential densities and more integrated land uses result in shorter work journeys, making walking and cycling possible. And until recently policies discouraged individual vehicle ownership.[1]

Cycling and walking were the dominant form of transportation as late as 1992, the year of the most recent survey of urban travel (table 6.2). Since then motorcycle use has increased dramatically, mostly at the expense of public transit and cycling. Increased urban sprawl—with longer trips, increasing incomes, and falling residential densities—will increase the reliance on motorized transport unless the quality of public transit improves.

TABLE 6.1
Motor vehicles account for a large share of pollutant emissions
(percent)

Location, year	Carbon monoxide	Hydro-carbons	Nitrogen dioxide	Particulate matter
Beijing, 1989	39	75	46	
Guangzhou, 1993	87		67	
Shanghai, 1994	37	85	15	
European Union, 1990	72	40	62	
Japan, 1989			40	39
Korea, Rep. of, 1991	99	96	81	18
United States, 1993	78	36	44	21[a]

a. PM-10.
Source: Walsh 1996.

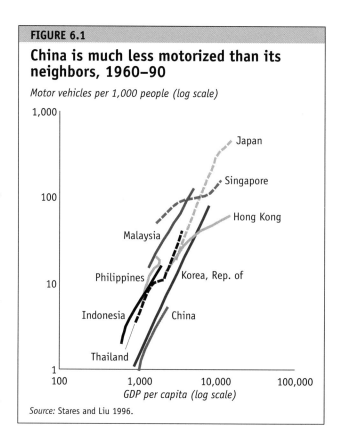

FIGURE 6.1
China is much less motorized than its neighbors, 1960–90

Motor vehicles per 1,000 people (log scale)

Source: Stares and Liu 1996.

TABLE 6.2

Walking and cycling account for sizable shares of urban transport

(percent)

Mode	Beijing (1986)	Shanghai (1986)	Guangzhou (1992)
Walk	13.8	38.2	30.6
Bicycle	54.0	34.2	33.8
Motorcycle	0.0	0.2	6.4
Taxi	0.3	0.2	6.1
Bus and metro	24.3	24.0	21.7
Car	4.4	2.2	0.0
Other	3.2	1.0	1.4
Total	100.0	100.0	100.0

Note: Data are for formal populations only.
Source: Li 1996.

Congestion and vehicle speeds

The slow operating speeds of motor vehicles in major Chinese cities result in high fuel consumption. Speeds on Beijing's third ring road average less than 20 kilometers an hour (Walsh 1996). At these low speeds, fuel consumption is twice that on more freely flowing urban highways. Even a small reduction in average speed—say, from 20 kilometers an hour to 15 kilometers an hour—increases fuel consumption by 25 percent (Stares and Liu 1996). Stop-and-go driving also substantially increases emissions (table 6.3). And congestion erodes the performance of public transit, such as buses and mini-buses.

Air quality standards and vehicle emissions

The National Environmental Protection Agency has issued eight motor vehicle emission control standards. Standards for light-duty vehicles, including cars, were adopted in 1983 and are similar to European standards from the same year. Environmental performance standards for heavy-duty vehicles were adopted in 1989 and for motorcycles in 1993.

Periodic inspections are required for all operational vehicles and are conducted by local vehicle management offices. The effectiveness of the tests varies significantly by jurisdiction.

The current vehicle fleet is highly polluting, in part because of outdated vehicle designs and inadequate emission standards for new vehicles (Fridley 1995). For example, truck engines are based on designs that are more than twenty years old, and most car engine

TABLE 6.3

Different driving modes generate different emissions, United States, 1984

(parts per million)

Driving mode	Hydrocarbons	Carbon monoxide	Nitrogen oxides
Idling	1.34	16.19	0.11
Cruising			
15 mph	5.11	67.36	0.75
30 mph	2.99	30.02	2.00
45 mph	2.90	27.79	4.21

Source: Faiz 1990.

designs are at least ten years old. Current emission standards for gasoline-fueled engines are based on 1978 European and 1981 U.S. standards. Vehicles do not require catalytic converters to meet current standards. Emission standards for diesel-fueled engines are also weak, covering only tailpipe smoke and not particulates and nitrogen oxides.

While emission standards in industrial countries have become increasingly stringent, those in China have stagnated. Chinese standards for cars allow forty times more carbon monoxide, six times as many hydrocarbons, and eight times as many nitrogen oxides than U.S. standards. Motorcycle standards are even more lax than those for cars, and standards for two-stroke engines (mopeds and three-wheelers) are similar to those for trucks.

But even these standards are weakly enforced. Emissions from in-use vehicles are ten to fifty times those of vehicles in the United States and Japan. Relative to U.S. automobiles, domestically manufactured Chinese cars emit thirty to forty times as much carbon monoxide, forty to sixty times as many hydrocarbons, and eight to fifteen times as many nitrogen oxides.

Fuels

Gasoline supply is split about evenly between 70 MON (equivalent to 82 RON) and 90 RON and higher grades. The lower (70 MON) grade, mostly unleaded, is used primarily in low-compression truck engines, while the higher grades are predominantly used in urban areas, where most higher-compression cars operate. Chinese leaded gasoline contains about 0.12 grams of lead per liter—less than the average of 0.15 grams per liter for other Asian countries but more than the previous international standard of 0.08 grams per liter, which is now considered high (World Bank 1995). Atmospheric lead levels in major cities such as Beijing typically are 1.0–1.5 micrograms per cubic meter ($\mu g/m^3$), more than twice the World Health Organization (WHO) standard, and in some areas reach 14–25 $\mu g/m^3$ (Walsh 1996).

Although about half the gasoline produced in China is unleaded, most of this is low grade (70 MON) or higher grades for export. The domestic supply of 90 RON and higher-grade unleaded gasoline accounts for only about 20 percent of gasoline consumption, and much of it is contaminated by leaded fuel at storage terminals. Wider use of unleaded gasoline is hindered by the lack of a dedicated storage and distribution system. Refineries are not responsible for distributing fuel, so most fuel is transported by truck or rail car rather than by pipeline. Since few of these vehicles transport only unleaded fuel and refueling sites generally have only one holding tank, unleaded gasoline is often contaminated with lead (Wang 1995). Even small amounts of lead inactivate catalytic converters.

The quality of China's diesel fuel is poor—with low stability and a high aromatic content—resulting in high vehicle emission of particulates and smoke. Although the sulfur content of nationally produced diesel is low by international standards, the average sulfur content of diesel fuel will increase in the future if the transport sector relies more on imported diesel (much of the imported diesel, which has up to 2 percent sulfur content, is currently used for power generation). Most East Asian countries limit the sulfur content of diesel to less than 1.5 percent.

Fuel pricing

The pump prices of gasoline and diesel are low by international standards (table 6.4). The current small price differential between 82 RON and 90 RON gasoline does not justify the cost of upgrading refineries to produce more high-quality fuel. Moreover, a shortage of liquefied petroleum gas is driving up its price, and refiners are diverting their feedstock to producing it rather than regular gasoline. In addition, since diesel is the main fuel for agricultural use, it remains underpriced relative to gasoline, reducing refinery incentives to produce diesel (Fridley 1995).

China's urban environmental future

Three sets of transportation-related factors will influence China's urban environmental future. The pace of income growth and urbanization will dictate the demand for urban transportation. Urban planning and infrastructure development will influence urban layout and the distance workers have to commute. And the supply options available for transport—bicycles, cars, or public transit—will provide competing choices. If demand exceeds supply, congestion will extend travel times, reduce average speeds, and increase emissions per trip.

Demand issues: Vehicle use and urban growth

China's rapid urbanization shows few signs of abating. By the end of 1995 the formal urban population stood at more than 350 million people and had been growing by 4 percent a year since 1980. The informal population was estimated at about 50 million and has been growing even faster. The urban share of the population is expected to reach 42 percent by 2010 and 50 percent by 2020, more than doubling the total number of urban residents.

Since 1990 the number of cities has been growing by almost 8 percent a year and the number of townships,

TABLE 6.4

China's fuel prices are low by international standards, second quarter 1997
(cents per liter)

Economy	Premium gasoline (unleaded where available)	Diesel
Asia		
China	25	22
Japan	97	65
Korea, Rep. of	94	41
Taiwan, China[a]	60	41
Thailand	37	34
Vietnam	40	31
Europe		
France	111	77
Germany	101	72
Italy	113	84
Spain	82	64
Americas		
Brazil	83	45
Canada	48	40
Mexico	38	30
United States	36	31

a. Data are for August 1996.
Source: World Bank data.

by 13 percent a year. The government plans to establish about 400 new cities over the next fifteen years.[2] Although this would create an average of twenty-eight cities a year, it is a significant reduction in the rate of increase of the past five years. The numbers of townships will need to increase even faster to cope with projected urban population growth and to avoid having large urban centers develop into megacities. The development of new cities and the expansion of smaller settlements into townships provides more scope for managing the transport impacts of urban growth than would the unconstrained growth of larger cities.

Until recently high population densities have helped constrain the demand for motorized transport in China. But the low quality of life resulting from these high densities is propelling their reduction, which will reduce the potential for nonmotorized travel and increase the need for road infrastructure. The planned reduction in residential densities in areas with densities of nearly 20,000 people per square kilometer will not create excessive demand for urban road space, but in new residential areas with densities of about 10,000 people per square kilometer the need for road space will be much higher (figure 6.2). The demand for motorized travel will also increase as land use becomes less integrated. The authorities plan to move people away from polluting industries, which will increase the separation of land uses.

The portion of built-up land area devoted to roads is low—less than 10 percent in major cities (Li and others 1996). Right-of-way on these roads is shared by pedestrians and nonmotorized and motorized vehicles. With poor traffic management and little segregation of users, speeds are low and accident rates are high. The lack of road space will become a critical issue as car ownership increases. To prevent traffic congestion from getting worse, road space should increase at about the same rate as the increase in the vehicle fleet. Achieving this objective will be difficult because of cost, resettlement, and space considerations, so congestion will probably only get worse.

Supply options: Car ownership

In the absence of good public transit options, the demand for private automobile ownership and use will rise. About one-third of China's vehicle fleet is passen-

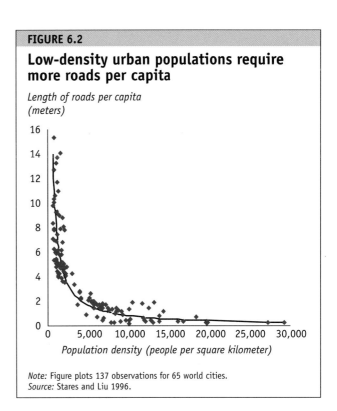

FIGURE 6.2

Low-density urban populations require more roads per capita

Length of roads per capita (meters)

Population density (people per square kilometer)

Note: Figure plots 137 observations for 65 world cities.
Source: Stares and Liu 1996.

ger vehicles (table 6.5). This segment of the fleet has grown nearly twice as rapidly as trucks. While most of these cars are owned by companies or government agencies (80 percent in 1995), privately owned cars are the most rapidly growing segment of the automobile market.

Future car ownership in China, particularly household car ownership, will depend on three main factors:
• Affordability (the ratio of car price to per capita income).
• Need for car use.
• Infrastructure available for car use.

Considering only affordability and the need for car use, and ignoring any effects of congestion or changes in public transit, a study sponsored by the State Planning Commission estimated that household car ownership would surpass the benchmark level of 15 cars per 1,000 people when the affordability ratio falls to 2–3 (table 6.6). Based on a random survey of nearly 10,000 high-income households, the study calculated the number of households that would be able to afford and be willing to own a private car and estimated car ownership, potential domestic car production, and fuel consumption.

Car ownership rates in urban areas are much higher than the national average, with about 24 cars per 1,000

TABLE 6.5
The registered civil vehicle fleet is growing quickly, 1985, 1990, and 1994
(thousands)

Type of vehicle/owner	1985	1990	1995	Average annual growth, 1985–95 (percent)
Passenger vehicles	795	1,622	4,179	18
Professional transport	83	108	131	5
Privately owned	19	241	786	45
Trucks	2,232	3,685	5,854	10
Professional transport	194	198	142	–3
Privately owned	265	573	1,226	17
Other	616	411	1,227	7
Total	3,642	5,717	11,260	13

Source: China Statistical Yearbook 1996.

people in Beijing, 15 per 1,000 in Shanghai, and 21 per 1,000 in Guangzhou (Stares and Liu 1996). These rates are low by international standards, even taking into account China's low per capita income, so the potential for growth is enormous. Most studies in other countries of the factors that affect car ownership find a strong correlation with personal income, and recent studies have found a close relationship between the quality of public transit and residential density (high densities being conducive to better bus and metro services).

Car ownership rates in some urban areas are expected to grow by four to seven times over the next fifteen years and could double again in the following ten years. Actual growth rates will depend on how residential densities change as existing cities grow and new cities are built and on the success of policies for improving mass transport. Cities with lower population densities and poor public

transit may have 20 percent more cars than the projected average. But cities that organize themselves well, with higher densities and good public transit, would have 20 percent fewer cars than average (table 6.7).

Supply options: Public transit

Efficient public transit is associated with high residential densities and results in less dependence on automobiles. So far, however, provision of good public transit has not been a policy priority. Bus services have suffered from underinvestment, inefficient management, poor maintenance, and a lack of user orientation, resulting in poorly maintained vehicles and unreliable services. Congested urban roads also result in unreliable services, with long delays between buses and unpredictable travel times. Lower service levels could cause ridership to fall off and increase the need to subsidize fares or further reduce services. Few cities have built underground railways, none has a surface suburban railway, and light rapid transit has yet to be used. Fixed-link public transit systems are now so expensive that they can be justified only in exceptional circumstances. Although the poor quality of public transit did not contribute to air pollution when the only alternative was to walk or travel by bicycle, the consequences are more significant now that average urban incomes are high enough to afford cheap motorcycles with heavily polluting two-stroke engines.

If China were to develop public transit options ahead of urban transport demand, major cities would have 20 percent fewer automobiles and, based on experience elsewhere, about 35 percent less car use.

TABLE 6.6
National car ownership and production should rise considerably, 2000, 2005, and 2010

Item	Projections		
	2000	2005	2010
Households able and willing to buy a car (millions)	4.0–4.7	15.5–16.5	37.5–40.5
Demand for cars (millions)	1.3–1.6	2.2–2.7	3.5–4.4
Household demand (millions)	0.4–0.6	1.2–1.6	2.3–3.0
Car ownership (millions)	6	12	22
Household car ownership (percent)	20	40	60
Cars per 1,000 people	4.7	9	15.8
Car production capacity (millions)	1.2	2.2	3.5
Car production capacity for household use (millions)	0.4	1.2	2.3
Household car gasoline consumption (millions of tons)	0.8	2.8	6.5
Share of national gasoline consumption (percent)	2	6	11
Average gasoline consumption (tons per household car)	0.7	0.6	0.5

Source: TEI 1994.

China's urban transportation future: The strategic choice

Unlike many developing countries, China has a strategic choice before it that will shape its urban environmental future. One option is to let market forces, rising incomes, and uncontrolled urban growth create cities. This option entails high social costs of lost worker productivity and damage to human health as well as rising demand for funds for road construction.

But there is a second option. China's unusually low dependence on automobiles, the unusually high density of its cities, and the strength of its administrative capacity at both the national and municipal levels is unique among developing countries. China can take advantage of these circumstances to create a new framework for urban transport and environmental management. In addition to developing and enforcing regulations and environmental standards, this option would require integrated actions in six areas: planning land use, supplying transport infrastructure, managing demand, managing traffic, promoting car production as a pillar industry, and implementing technical standards for vehicles and fuels.

Land use planning

China can take advantage of its unique pattern of urban development (Li and others 1996). Instead of dispersing high-density populations, urban growth could be organized along corridors with established public transit services at the transport nodes along those corridors. This approach would minimize automobile use since it would be possible to walk or cycle to the transport nodes. Commuters traveling further could transfer to the high-quality public transit services.

Curitiba, Brazil, and Portland, Oregon (United States), are among the few cities that have successfully implemented this model since the advent of high car ownership and use, although it was the basic pattern of urban growth in the nineteenth century (box 6.1). For maximum effect, the policy needs to be introduced before urban growth is too far advanced, so that the land for public transit services and radial roads can be made available without excessive relocation of residents and high land costs. China's current strategy of creating new townships and cities and limiting the growth of existing large cities will support the application of this strategy.

Because this strategy is not as easy to apply in existing cities, a more dramatic solution, based on demand management coupled with excellent public transit and the development of regional centers, will minimize the demand for radial travel and ease the pressure for more road construction. This second strategy can also reduce the demand for car ownership, as it has in Hong Kong (China) and Singapore (see figure 6.1).

Public investments and the supply of infrastructure

The demand for urban road construction will probably exceed the supply for some time. The experience of other countries suggests that this demand will quickly find its way into automobile purchases, but soon traffic jams will put pressure on the public sector to build more roads (table 6.8). But if China relies on an automobile-

TABLE 6.7

By 2010 urban car ownership will be considerably higher
(cars per 1,000 people)

Density and public transit assumptions	Beijing	Shanghai	Guangzhou	Chengdu	Jinan
Current ownership rate	24	15	21	8	15
Basic projected rate[a]	96	105	116	55	70
Actual residential density[a]					
Better public transit	77	84	92	50	58
Worse public transit	106	116	127	65	80
Reduced residential density[a]					
Better public transit	92	100	110	65	69
Worse public transit	116	127	140	87	90

a.Data are projections.
Source: World Bank staff estimates.

A model for China's urban development?

Curitiba, Brazil, a city of 2 million people and an area of 1,100 square kilometers, has managed to control the twin challenges of rapid urbanization and motorization. In the late 1960s city officials feared that immigration and rapid growth would overwhelm the city, ruin the downtown, and turn the city into an uncontrollable, sprawling metropolis—as had occurred in other Brazilian cities.

City planners adopted comprehensive transportation and land use planning to manage the surge of new entrants to the city. Mindful of budget constraints, they required simple, low-cost solutions that could be implemented quickly and easily, creating momentum and political credibility. Development focused on five linear corridors and coordinated transport and land use planning to shape the new metropolis. The primary tools were a system of exclusive and integrated busways, a transit-supportive zoning law, and restrictions on automobile use in the downtown area. The privately operated all-bus transit system today offers a rich and integrated variety of services, carrying more than 1 million passengers a day and accounting for 75 percent of work commutes. This outcome has been achieved in a relatively rich city (per capita GDP was $5,149 in 1990) with high rates of car ownership (267 cars per 1,000 people).

Convenient transit services result in car use and gasoline consumption 30 percent below those in comparable Brazilian cities, helping to produce a living environment with low vehicle emission and congestion levels. With more than twenty years of development, Curitiba's model has proven highly sustainable.

The main lessons of Curitiba's experience? Urban planning must be comprehensive and make future transportation needs its centerpiece. In addition, city officials relied on the local Urban Planning Institute, which provided high-quality, innovative solutions to planning problems. The institute, with the support of the municipal government, worked closely with businesses and communities in developing ideas and planning concepts. As a result community ownership and pride in city achievements became an integral part of the city development process.

Source: World Bank staff based on Cevero 1995.

based urban development strategy, public investments in roads will never catch up with private investments in automobiles.

The costs of an automobile-based strategy are much higher than those of a public transit strategy. The following estimates show the total investment needed in urban roads and public transit systems based on two assumptions—one of no change from current policies and the other of an emphasis on public transit corridors and associated lower projected levels of car ownership. Both estimates are less than that projected for interurban transport (highways, railways, ports, and airports) and represent about 2 percent of GDP; the public transit option would be about 17 percent less costly than the automobile-oriented strategy.

In both strategies highway maintenance would be the largest expenditure, accounting for 30–44 percent of the total. In the automobile-based strategy new ring roads would account for a further 30 percent, but this would only be 12 percent under the public transit strategy.

Many Chinese cities plan to build ring roads to expressway standards, sometimes with radial routes connecting them to the downtown business area. In cities in the rest of the world, however, the share of road space dedicated to ring roads is quite small and varies within a narrow range.[3] Current plans for expressway construction go well beyond the standard in developed motorized cities around the world.

Why have motorized cities in other parts of the world stopped short of such elaborate expressway plans? In part, because of the huge costs of constructing urban expressways, especially in city centers (where land is costly). Because the intrusion of these roads disrupts communities and encounters local resistance, few new expressways have been built in Western city centers

TABLE 6.8
Public spending on roads will eat up a large share of investment in urban transport, 2000–10
(billions of U.S. dollars)

Investment	Automobile-based strategy		Public transit strategy	
	Annual investment	Share of total	Annual investment	Share of total
New roads	14.9	51.2	8.9	36.9
Ring roads	8.5	29.2	3.0	12.4
Arterial roads	3.5	12.0	3.5	14.5
Other roads	2.9	10.0	2.4	10.0
Highway maintenance	12.8	44.0	9.4	30.0
Buses	1.4	4.8	4.0	16.6
Metros and light rapid transit	0.0	0.0	1.8	7.5
Total	29.1	100.0	24.1	100.0

Note: Data are projections.
Source: World Bank staff estimates.

in recent years. Moreover, the notion that it is possible to build a way out of congestion is being increasingly questioned.

Urban transport investment in China should average about $25 a year billion over the next twenty years to avoid further deterioration in travel speeds and congestion-induced pollution. A substantial portion of the total should be allocated to road maintenance and improvements to existing arterial roads, with a reduction in planned expenditure on ring roads. Investment in fixed-link public transit should be restricted to corridors with high demand and areas where it is cost-effective to avoid building new highways. Investment in buses should be aimed at improving service quality and frequency so that commuters are not tempted to turn to cars.

Demand management

Demand management should be aimed at making the best use of existing infrastructure, incorporating environmental costs, and responding to the demand for new facilities. It can also be used to reduce the unit costs of public transit by reducing peak demand and stimulating off-peak use. The tools of demand management combine pricing and regulatory measures designed to make people's travel choices take account of the economic consequences of their actions. They aim to manipulate demand to avoid both excess use and overloaded infrastructure on the one hand and unnecessary duplication and underuse of capacity on the other. While fixed toll systems can be used effectively on some principal roads, they are not an efficient way of managing urban transport demand. New technology will make electronic charging for road use more feasible. Given that electronic charging provides the opportunity for an accurate and sensitive road pricing system and that China's cities are embarking on a major period of new road building, incorporating these technologies would not add excessively to the cost of road development.

Demand management will need to supplement investment to create an efficient urban transport system. Large cities should introduce appropriate pricing and regulatory measures to restrict car and motorcycle use on congested roads or in congested areas, minimize the cost of public transit services, and reduce congestion on main highways.

Traffic management

Traffic management is designed to make the best use of space to accommodate demand with minimal investments in new infrastructure. Using traffic engineering, traffic control, and traffic enforcement to manage traffic flow, traffic management gives priority to the movement of people rather than vehicles. Street capacity in China can be increased by installing junction improvements and traffic signals, creating bus lanes, limiting and otherwise controlling on-street parking, and introducing one-way streets and other traffic flow measures.

One of the most cost-effective traffic management measures is to time traffic signals according to the flow of traffic or to the priority desired for high-occupancy vehicles such as buses. Timed signals can increase total highway capacity by up to 20 percent. Although a few Chinese cities have invested in such systems, they are not as effective as they should be and are not designed to give priority to buses.

In many parts of the world an initial response to urban traffic congestion has been to require downtown offices to provide adequate off-street parking so that on-street parking controls can be enforced. Over time, however, this move often induces additional car travel, ultimately causing both on-street and off-street parking to be severely restricted. Depending on the stage of development at which the change is made, its effectiveness is limited by the amount of available off-street parking. Thus municipalities in China should resist pressures to make off-street parking a condition of approving a building license, and go straight to the second stage of making a low level of off-street parking a condition of approval. But this policy must be accompanied by one that makes other modes of transport—walking, cycling, and public transit—attractive alternatives; otherwise it will be impossible to impose on-street parking controls. The use of bicycles in cities requires special traffic management techniques in order to protect their use and avoid conflicts with motorized traffic.

No traffic management system can function if it is not enforced. In China the responsibility for designing and implementing traffic management is shared by municipalities, the Ministry of Construction, and the Public Security Bureau, but the allocation of responsibility among them is not always efficient. Given the importance of traffic management in adding to system capacity and reducing congestion, these responsibilities should be clarified and the various measures vigorously applied.

Area traffic management systems should be developed and implemented to deal with the mix of motorized and nonmotorized traffic in Chinese cities. Where their advantages can be demonstrated, systems that give priority to buses and other high-occupancy vehicles should be designed and implemented. Planning standards for new commercial developments should include reference to the development's provision of parking spaces, and where adequate public transit is available, the number of parking spaces should be strictly limited. Regulation of on-street parking and observation of traffic markings and signals should be better coordinated between the ministry, the bureau, and municipal governments.

Traffic accidents in urban areas, already high by international standards, are increasing at alarming rates throughout China. Congestion generates impatience and a deterioration in driver discipline and driving conditions. Road networks in many cities were never designed for current traffic volumes and speeds. The push to provide more road space for motorized vehicles has resulted in inadequate facilities for pedestrians and slow-moving vehicles. The government must improve road safety by raising road design standards, imposing road safety legislation and penalties, enforcing traffic laws, testing and inspecting vehicles, strengthening driver training and testing, and educating children and adults in traffic safety.

Car production as a pillar industry

Exercising control over automobile ownership and use is not necessarily inconsistent with China's objectives of developing its automobile industry. The potential size of China's domestic car market is so huge that it will not require artificial stimulus to promote an export industry. Other Asian countries with successful car export industries, such as Japan and the Republic of Korea, have already implemented policies to invest in public transit and contain the demand for car use. National environmental policy recognizes that the development of cars will be controlled to a "proper extent and consistent with the actual situation in China" and that a

better solution is to focus on developing various forms of public transit (NEPA 1996).

In the meantime, however, extending the "polluter pays" principle to automobiles and other vehicles would require companies and individuals using cars to pay for the social costs of pollution. Thus motorcycle and car owners would be faced with the full costs of their choices, resulting in economically efficient outcomes. The revenues from the charges could be used to finance additional pollution and congestion mitigation measures.

Policies in several Asian cities make car ownership expensive. Singapore's example is well known, but Hong Kong (China) and Seoul (Korea) also have high car purchase and ownership taxes. Korea is a particularly interesting case because price disincentives to operate cars are used in parallel with a successful policy of developing an export market for cars. This is a good example for China, where the potential domestic market would still be huge even with such taxes in place.

Japan distinguishes between ownership and usage. Despite high car ownership, many Japanese cities are relatively free of congestion because of low car usage. These cities have well-developed, efficient public transit services and strict controls on parking. In addition, parking is prohibitively expensive in downtown areas and there are few off-street parking spaces. Nearly 90 percent of motorized trips in Tokyo are by rail-based public transit.

Vehicle emissions and fuel standards

Three policies can be used to modify the air quality impact of increased motorization: imposing emission standards on new vehicles, affecting their design; controlling emissions from in-use vehicles; and setting the prices of the fuels that vehicles consume. A start has been made on addressing vehicle-induced pollution at the national level through vehicle registrations and ownership regulations, but these are weakly enforced.

Regulation of fuel quality is fragmented, with jurisdiction over fuel quality shared among several agencies, including the China Petrochemical Corporation (industry standards) and the National Technical Supervision Bureau (national standards on petroleum products and other products). While the bureau has final authority for setting standards, the corporation, which produces about 87 percent of China's motor fuels, is responsible for proposing those standards. As with other standards in China, this approach has resulted in standards that match industry needs more than environmental needs. The National Environmental Protection Agency, which sets national standards for lead, sulfur dioxide, particulates, nitrogen oxides, carbon monoxide, and hydrocarbons, historically has had little input into the process.

Pollution from motorized vehicles comes from the fuels they consume, so an appropriate system of pollution charges would relate to fuel prices. In addition to a market price based on a sum of the opportunity cost of the fuel (equal to its international market price), the cost of distribution, and ad valorem taxes on consumption (as for other products), the fuel price could include an infrastructure surcharge and a pollution charge. The pollution charge could be proportional to the energy content of the fuel and its potential for pollution, including its sulfur and lead content.

Emissions testing of new vehicle models should be carried out by an agency that is independent of vehicle manufacturers and is responsible for air quality, such as the National Environmental Protection Agency. Emission standards should be revised more frequently, giving manufacturers incentives to improve the emissions performance of their products. Once a system for distributing unleaded gasoline is in place, much stricter standards requiring the use of catalytic converters should be introduced. Testing of in-use vehicles needs to be more stringent and uniform, at least in urban areas. Similarly, fuel standards should not be left entirely to the petroleum industry and the China Petrochemical Corporation to determine, but should be the responsibility of the National Environmental Protection Agency. Transport fuels should include a surcharge according to their polluting capacity, to provide an incentive to use less damaging fuels and promote the more efficient use of all fuels.

Technical standards for transport fuels

China plans to phase out lead by 2000. The experiences of other countries suggest that this will be a challenging task. Other planned improvements in fuel quality include eliminating low-octane gasoline (70 MON production), raising the octane level by increasing the use of reformate, oxygenates, alkylates, and other high-octane blending components, and using detergents to control deposits in gasoline-fueled engines.[4] Other low-polluting hydrocarbon fuels used for transport in other countries include liquefied petroleum gas, liquefied natural gas, methanol, ethanol, and biogas. However, the feasibility of these alternatives usually depends on a readily available supply, often as a byproduct of the production of other hydrocarbon products. China will not have an excess supply of any of these fuels in the near future, so they are unlikely to play a significant role in reducing vehicle emissions.

Widespread use of unleaded fuel will permit the introduction of new environmental standards that can only be met by requiring motor vehicles to be equipped with catalytic converters. Although the rest of the world uses vanadium and similar rare earths as the catalyst, China has an abundance of lower-priced alternative minerals that could function almost as well, so the increase in initial vehicle costs will be negligible. Given the time it will take to develop a parallel distribution system for unleaded fuel, unleaded fuels should be phased into use. They should first be made available in those urban areas where lead contamination is greatest.

Lowering lead emissions will have high and immediate payoffs to society by reducing the incidence of lead poisoning among millions of Chinese children. Although the complete phaseout of lead will be difficult and costly—requiring significant industry coordination and investment—it will be even more costly if it is left until vehicle use and fuel consumption are much higher. Some refineries are already producing higher-octane unleaded fuel (part of the output is exported), but the greater cost will be developing a parallel distribution system, essential so long as both leaded and unleaded fuel are on the market. In the United States cooperation between the automobile and oil industries resulted in nationwide availability of unleaded gasoline in four years, but it took about ten years to completely eliminate lead from the fuel supply.

The market pricing of unleaded gasoline must be carefully controlled. If it is priced cheaper than leaded gasoline while its supplies are limited, there is a risk that it will be used by vehicles that do not need it (that is, those not equipped with catalytic converters), leaving insufficient supply for those that cannot operate without it. Owners of catalytic converter–equipped vehicles could then be tempted to use leaded gasoline, destroying their catalytic converters and eliminating any air pollution reduction. But if unleaded gasoline is more expensive than leaded gasoline, owners of cars equipped with catalytic converters will be tempted to use leaded gasoline, with the same final result. Thus, at least until adequate supplies of unleaded gasoline are available, there are good reasons not to differentiate between the price of leaded and unleaded gasoline.

Since domestically produced diesel is already low in sulfur, it should not be difficult to control sulfur dioxide emissions from diesel-fueled vehicles. But it will be difficult and expensive to achieve the particulate emission standards of some other countries because of the low stability and high aromatic content of domestically produced diesel.

It is difficult to justify an increase in the price of diesel for transport purposes (to encourage more efficient use) since more diesel is used in agricultural tractors than in road vehicles. There is a possibility of having two classes of diesel fuel, one for transport and the other for everything else, with the second retaining its low price. Such a policy would require strict enforcement to avoid transport users simply buying the cheaper alternative. This policy is applied in the United Kingdom with some success.

Air quality effects of alternative urban transport strategies

Three scenarios were developed to simulate the potential effects of different urban transport strategies on fuel demand and pollutant emissions. The first scenario assumes that the current policy framework remains unchanged. The second and third scenarios assume that all other measures to reduce vehicle emissions, particu-

larly introducing unleaded gasoline and low-sulfur diesel and implementing higher emission standards, will be taken. Lead emissions will only fall to zero when all the vehicles that depend on leaded fuel have been withdrawn from the vehicle fleet and distribution of even low-leaded fuel has stopped.

If there is no change in urban growth trends—that is, if urban densities continue to fall, investment in public transit remains inadequate, demand and traffic management are avoided, and there is no improvement in vehicle emission standards and their implementation—exhaust emissions will be about twenty-five times higher by 2020 (table 6.9). Implementing tougher emission standards and improving traffic management would hold the level of increase to about fifteen times. If a bus-based urban public transit system were promoted as well, vehicle emissions would be only about five times higher.

A public transit strategy would also lower future oil consumption by two-thirds. At current prices, in 2020 the public transit strategy would require only $30 billion in gasoline and diesel consumption, compared with $87 billion under an automobile-based strategy.

In summary, if China wants to prevent skyrocketing oil demand from automobile use and effectively control urban air pollution from automobile operation, it must take early and decisive action to plan cities around viable public transit systems. Such actions will also free up petroleum and capital for other purposes—for example, as a substitute for coal in residential and commercial energy use—thus creating ripples of environmental benefits.

The larger task of creating a clean and healthy urban environment will require harnessing China's considerable administrative and coordinating capacity in transportation management. Emission controls and cleaner fuels are

TABLE 6.9
A public transit strategy would mitigate increasing fuel demand and worsening pollutant emissions

Fuel or pollutant	Base year, 1993	Automobile-based strategy without improvement, 2020	Automobile-based strategy with improvement, 2020	Public transit strategy, 2020
Gasoline (millions of tons)	29	789	473	144
Diesel (millions of tons)	8	144	86	49
Carbon monoxide (thousands of tons)	8,389	215,359	55,105	15,831
Lead (thousands of tons)	4.6	126	0	0
Nitrogen oxide (thousands of tons)	956	22,305	8,824	3,094
Volatile organic compounds (thousands of tons)	1,138	22,895	6,025	1,948
Particulate matter (thousands of tons)	80	1,838	868	362

Source: World Bank staff estimates.

essential. Achieving them will require greater municipal planning to shape land use patterns in economically efficient ways, greater investment in coordinating mass transport, and greater efforts to accommodate bicycles and other nonmotorized uses. The strategic choice is clear.

Notes

1. In 1994 the government affirmed its policy of making automobile manufacturing a pillar industry and began encouraging private ownership of cars.

2. Many of these new cities will be upgraded towns.

3. Several East Asian cities with high levels of motorization have similar shares of road space dedicated to ring roads: 2.4 percent in Tokyo (Japan), 2.6 percent in Seoul (Korea), and 3.4 percent in Singapore. But an average of sixteen cities throughout the world has an average of just 1.7 percent. Even U.S. cities have an average of less than 3 percent with a similarly small standard deviation, considering that the sample embraces a wide range of city types, from New York to Los Angeles. If Chinese cities follow this pattern, the entire expressway "allocation" for most cities might be taken up by just one ring road and one or two radial routes, with perhaps an additional ring in the largest cities (Stares and Liu 1996).

4. The China Petrochemical Corporation, China's main motor fuel producer, plans to upgrade the quality of gasoline using advanced refining technology and alkylates and oxygenates in order to reduce vehicle emissions (Wang 1995).

References

Cevero, Robert. 1995. "Creating a Linear City with a Surface Metro: The Story of Curitiba, Brazil." University of California at Berkeley, National Transit Access Center.

China State Statistical Bureau. 1995. *China Statistical Yearbook 1995*. Beijing.

Faiz, Asif. 1990. "Automotive Air Pollution: Issues and Options for Developing Countries." Policy Research Working Paper 492. World Bank, Washington, D.C.

Fridley, David. 1995. "China Fuel Quality Specifications." World Bank, Washington, D.C.

Li, Yaming. 1996. "Annex: Urban Statistics and Transport Characteristics." In Stephen Stares and Zhi Liu, eds., *China's Urban Transport Development Strategy*. Washington, D.C.: World Bank.

Li, Yaming, and others. 1996. "Land Use and Transport Planning in China." In Stephen Stares and Zhi Liu, eds., *China's Urban Transport Development Strategy*. Washington, D.C.: World Bank.

NEPA (National Environmental Protection Agency). 1996. "Seclected Documents from the Fourth National Environmental Conference." Beijing.

Stares, Stephen, and Zhi Liu. 1995. "Motorization in Chinese Cities: Issues and Actions." In Stephen Stares and Zhi Liu, eds., *China's Urban Transport Development Strategy*. Washington, D.C.: World Bank.

TEI (Technical and Economic Institute of the State Planning Commission). 1994. "The Development Strategy for Cars to be Used by Households in China." *Economic Daily*, October 24.

Walsh, Michael P. 1996. "Motor Vehicle Pollution Control in China: An Urban Challenge." In Stephen Stares and Zhi Liu, eds., *China's Urban Transport Development Strategy*. Washington, D.C.: World Bank.

Wang, Xichu. 1995. "The Features of Motor Fuels and Lubricants in China and their Future Directions." Research Institute of Petroleum Processing, China.

World Bank. 1995. "Why Lead Should be Removed from Gasoline." Environment Department Dissemination Note 32. Washington, D.C.

chapter seven

Water: Preserving Scarce Resources

China has a long history of managing water resources to compensate for its uneven spatial and temporal distributions of water. In recent decades rapid economic growth and urbanization have brought new challenges to water management. Increasing soil erosion and reclamation of natural water reservoirs have heightened pressures for flood control. Rising water demand has outpaced municipal water supply capacity. Widespread water pollution has become a prominent environmental problem. And investment and reform in the water sector have lagged behind the general economy.

The underinvested and overburdened water system is taking its toll. Floods have become more frequent and more damaging. In 1996 floods caused 220 billion yuan in damages, compared with 160 billion yuan in 1994 (Chen 1996). More than 300 of China's 600 cities experience water shortages totaling about 6 billion cubic meters a year and causing 120 billion yuan a year in lost industrial output (SPC 1995). China's rivers, lakes, and reservoirs are all polluted to some degree. Except for some inland rivers and large reservoirs, water pollution has worsened, especially in waters adjacent to industrial cities and towns (NEPA, *China Environmental Yearbook 1995*). Water pollution causes annual damages on the order of 33 billion yuan (NEPA 1997).

While floods cause enormous damage, water shortages and water pollution are considered the two main water-related constraints to China's socioeconomic development, and are the focus of this chapter. Resolving both problems will require large capital investment and, equally important, effective policies and strong institutions.

Water quantity

China is relatively poor in water resources.[1] Among several major nations, China has the second lowest per capita water resources (behind India)—less than one-third the world average (table 7.1). Northern China, however, has only one-fifth the per capita water resources of southern China.

China is divided into nine water resource regions, roughly based on the boundaries of major river basins (table 7.2). Southern China (regions 5, 6, 7, and 8) contains 81 percent of China's water resources and 55 percent of its population. Northern China (regions 1, 2, 3, and 4) contains 14 percent of China's water resources yet must support 43 percent of its population. Northern China's 750 cubic meters per capita water resource

TABLE 7.1
China's water resources are limited—especially in the north, 1995
(cubic meters)

Country	Annual renewable resources	Water per land area (per square kilometer)	Water per capita
Canada	2,901	290,944	98,462
Brazil	6,950	816,494	42,957
Russian Federation	4,498	263,426	30,599
United States	2,478	264,659	9,413
China	2,812	292,917	2,292
Northern China	405	150,528	750
Southern China	2,278	654,532	3,440
India	2,085	700,840	2,228
World	41,022	301,988	7,176

Source: WRI 1996; SSB 1996.

average is just 10 percent of the world average. The lack of water is particularly acute in the Hai River Basin, which covers all of Hebei Province and Beijing and Tianjin municipalities and parts of Shanxi, Henan, Shandong, Jiangsu, and Anhui provinces. Groundwater in the Hai River Basin is overextracted by as much as 30 percent and in many areas groundwater tables have dropped by 100–300 meters. The depletion of indigenous water resources in the north is the primary reason for China's south-north water transfer project.

The temporal disparity of rainfall aggravates water shortages and causes devastating floods and droughts in major river basins. For example, 70–80 percent of rainfall in the Hai River Basin occurs between June and August. Such a pattern requires extensive water storage to ensure a stable supply.

Demand patterns and projections

In 1980 (the most recent year for which national data are available) irrigation was the main user of freshwater (82 percent), followed by industry (10 percent), rural activities (including household use, animal feeding, and nonfarming activities; 6 percent), and municipal use (2 percent; table 7.3). There is considerable variation among regions, with irrigation's share as high as 95 percent in the mostly rural northwest and as low as 69 percent in the highly industrial northeast. Over the past fifteen years the demand for nonirrigation water has increased quickly. For example, in the Hai River Basin irrigation's share of the total dropped from 81 percent in 1980 to 74 percent in 1987 (SPC 1995).

Total water demand is projected to reach 765 billion cubic meters in 2000, an increase of 72 percent over 1980. Nonirrigation uses will account for 56 percent of the increase in demand, with industrial and municipal demand expected to grow the fastest (see table 7.3). About 62 percent of the increase in water demand will come from the south, 35 percent from the north, and 4 percent from the inland rivers region.

China's maximum freshwater withdrawal is estimated at 1,100 billion cubic meters. At the national level the balance between demand and supply could be maintained. Striking this balance, however, will require huge investments in water supply and wastewater collection and treatment facilities, as well as enormous conservation efforts. Otherwise water shortages and

TABLE 7.2

Southern China contains half the population but most water resources
(percent)

| Region | Water resources | | Cultivated land | Population |
	Share	Per capita (cubic meters)		
Total	100.0	2,292	100.0	100.0
1. Northeastern rivers	6.9	1,625	19.8	9.8
2. Hai River	1.5	355	10.9	9.8
3. Huai/Shangdong rivers	3.4	515	14.9	15.4
4. Yellow River	2.6	749	12.7	8.2
5. Yangtze River	34.2	2,280	24.0	34.8
6. Southern rivers	16.8	3,534	6.8	11.0
7. Southeastern rivers	9.2	2,892	3.4	7.4
8. Southwestern rivers	20.8	32,216	1.7	1.5
9. Inland rivers	4.6	5,126	5.8	2.1

Note: These data were compiled in the early 1980s. It is assumed that the patterns have not changed. The 1995 population was used as the base for per capita water resource calculations.
Source: Wu 1989.

pollution will hamper economic growth and impair the quality of life.

Main causes of water shortages

Water shortages are a chronic problem in China. In 1980 estimated national shortages were already nearly 39 billion cubic meters, assuming average annual rainfall (Wu 1989). A more recent estimate puts annual shortages at about 60 billion cubic meters (Chen 1996). Increased shortages are as much a supply-side problem as a demand-side one.

In much of northern China water extraction is near or exceeds sustainable levels. Underinvestment in water supply infrastructure has caused water shortages in many urban areas that have adequate water resources. Increasing pollution has rendered many raw water sources unsafe, requiring expensive and disruptive relocation of these sources. A study of 270 cities with serious water shortages found that 136 have inadequate water resources, 110 have insufficient supply infrastructure, and 74 are short of water because water pollution has contaminated sources of raw water (SPC 1995).[2]

The rapid growth in demand is not entirely justified because water has not been allocated and used efficiently. Agriculture, the main consumer of water, wastes about 60 percent of the water passing through its old-fashioned ditch irrigation system. The expansion of rice planting in water-short northern regions has aggravated shortages (Lu 1993). Most industrial enterprises, many extracting their own water from rivers or the ground, do not recycle water. (The national industrial water recycling rate is about 40 percent, compared with about 70 percent in OECD countries.) Outdated production technologies also contribute to excessive water con-

TABLE 7.3

Water demand is rising sharply
(billions of cubic meters)

| Region | Municipal | | Industry | | Irrigation | | Rural | | Total | |
	1980	2000	1980	2000	1980	2000	1980	2000	1980	2000
Total	6.8	29.4	45.7	177.3	365.6	506.4	25.6	51.7	443.7	764.8
Northern China	3.2	11.1	17.9	69.3	132.2	171.7	9.5	22.0	162.8	274.1
Hai River Basin	1.1	3.0	4.9	19.0	30.9	36.2	1.5	3.7	38.4	61.9
Southern China	3.4	17.8	27.1	105.3	180.2	272.4	14.4	27.1	225.1	422.6
Inland rivers	0.2	0.5	0.7	2.7	53.2	62.3	1.7	2.6	55.8	68.1

Note: Data for 2000 are projections.
Source: Wu 1989; World Bank staff estimates.

sumption. For example, coal-based ammonia manufacturing consumes 500–1,000 tons of water per ton of ammonia, compared with 12 tons of water in the natural gas–based process. Chinese paper industries consume 400–500 tons of water to produce one ton of paper product, compared with 5–200 tons in OECD countries (NEPA 1997).

Solving the problems of water supply and demand will require concerted efforts in pricing, investment, regulation, and institutional reform.

Water quality

Water quality is largely determined by pollutant discharges. As noted, most of China's bodies of water are becoming increasingly contaminated by these discharges.

Pollutant discharges

The main sources of China's water pollution are industrial and municipal wastewater discharges, agricultural runoffs of chemical fertilizers, pesticides, and animal manure, and the leaching of solid waste.

In 1995 regulated industries discharged 27 billion cubic meters of wastewater into open bodies of water (figure 7.1). Municipal sources discharged 15 billion cubic meters. Although wastewater discharges from regulated industries have stabilized since the early 1990s, discharges from township and village industrial enter-

prises and municipal sources have increased rapidly, underscoring the urgency of controlling these enterprises' pollution and treating municipal sewage.

The relative pollution load is much greater in the water-short north because of its limited natural assimilative (dilution) capacity. In 1995 wastewater accounted for about 4 percent of surface water runoff in the north and 1 percent in the south, even though the south generates 64 percent more wastewater. In the Hai River Basin (water region 2) the wastewater-runoff ratio is a high 11 percent (figure 7.2). Overall, water pollution is much more severe in the north than in the south.

Among regulated enterprises, in 1995 chemical oxygen demand discharges were highest from paper (42 percent of industrial discharges), food and beverage (28 percent), and chemical (9 percent) industries. Petroleum pollutants were highest from chemical (29 percent of industrial discharges), metallurgical (21 percent), and machinery (13 percent) industries. The chemical industry is also the largest source of mercury (42 percent of industrial discharges), arsenic (46 percent), cyanide (42 percent), and volatile phenol (28 percent) discharges.

Water contamination from agricultural runoff is becoming more serious because of more intensive use of chemical fertilizers and pesticides as well as a huge increase in livestock production. Over the past decade chemical fertilizer applications doubled, reaching 36 million tons in 1995. Pesticide applications also doubled, to 1 million tons.

FIGURE 7.1

Wastewater discharges are on the rise

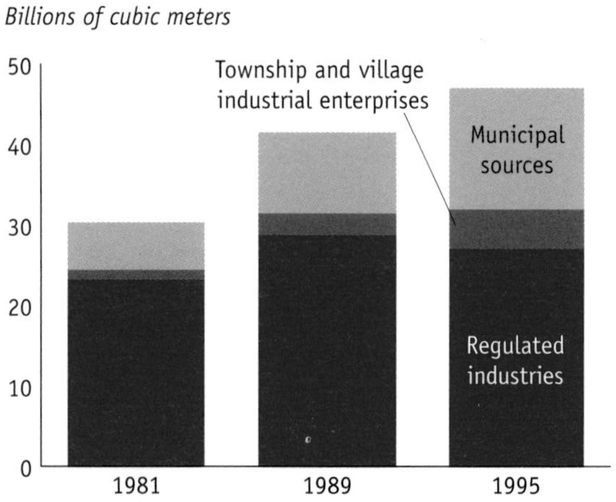

Billions of cubic meters

Township and village industrial enterprises

Municipal sources

Regulated industries

1981 1989 1995

Note: Wastewater discharges from township and village industrial enterprises in 1981 are estimated based on the 1989–95 increase.
Source: NEPA, China Environmental Yearbook 1996.

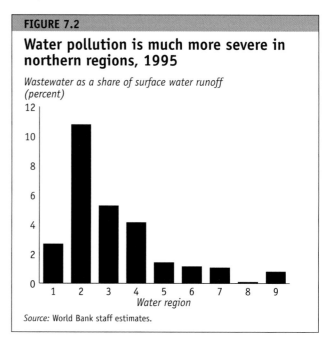

FIGURE 7.2

Water pollution is much more severe in northern regions, 1995

Wastewater as a share of surface water runoff (percent)

Water region

Source: World Bank staff estimates.

The leaching of solid waste, especially toxic waste, also may be an important source of water contamination. China has accumulated an enormous amount of solid waste—industrial solid waste now totals more than 6.6 billion tons, and is increasing by 180 million tons a year. About 10 percent of industrial solid waste is hazardous or toxic. In 1995 regulated enterprises alone dumped 6.5 million tons of solid waste into water bodies (NEPA 1996).

Mobile sources such as barges and ships also contribute to the contamination of rivers, lakes, and coastal waters through oil leaks and spills. Some 300,000 vessels discharge or leak more than 5,000 tons of oil into the Yangtze River each year. Wastewater discharges from these vessels into the Yangtze are estimated at 90 million tons a year. Solid waste discharges are estimated at 80,000 tons a year (MOT 1997).

Ambient water quality

The government has defined five categories of ambient freshwater quality standards. Grades 1, 2, and 3 permit direct human contact and use as raw water for potable water systems. Grade 4 is restricted to industrial use and recreational use other than swimming. Grade 5 is restricted to irrigation. Each grade specifies an acidity level (pH) and maximum concentrations for twenty-eight major pollutants. Exceeding the pH or any of the concentration standards for a given grade disqualifies the measured water body from being designated as that grade.

Urban bodies of water are among the most polluted because they receive large amounts of untreated industrial and municipal wastewater. National monitoring of 135 urban river sections found that 34 of 66 urban river sections in the north and 18 of 69 in the south are polluted beyond grade 5 standards, rendering them waste sinks (table 7.4). None of the sections was rated grade 1.

Urban sections of rivers are polluted mainly by organic matter, especially petroleum and permanganate. Other major pollution includes biological oxygen demand, ammonia nitrogen, volatile phenols, and mercury (table 7.5). Grade 3 standards are the minimum requirements for raw water sources for potable water systems. While most pollutant levels remained stable or fell slightly between 1991 and 1995, ammonia nitrogen increased by about 50 percent. This jump may be the result of increased discharges of untreated municipal sewage, agricultural runoff of nitrogen fertilizers, and discharges from fertilizer plants.

Major river systems in northern China are severely polluted, while those in the south (Yangtze and Pearl rivers) are relatively clean. Half of the cross-section water samples drawn from the Hai and Luan rivers do not meet grade 5 standards (figure 7.3). About 94 percent of the Daliao River is unsuitable as a raw water source for potable water. By 1995 industrial water pollution in the Huai River Basin was nearly disaster level, prompting the central government to adopt drastic measures to control wastewater discharges along the river.

Lakes and reservoirs generally suffer from eutrophication (abnormal growth of aquatic plant life caused by adverse biological enrichment) caused by increasing concentrations of nitrogen and phosphorus. Among China's large freshwater lakes, Taihu, Chaohu, and Dianchi are the most polluted. Five-year (1991–95) average nitrogen levels in the three lakes were 2.5, 3.8, and 2.3 milligrams a liter compared with the grade 5 standard of 2.0 milligrams a liter. Mercury concentrations are also high in the three lakes. The five-year average mercury levels of Chaohu and Dianchi exceed the grade 3 standard of 0.0001 milligrams a liter (NEPA 1996).

Groundwater quality has been declining, especially in the north, where groundwater is used intensively to compensate for the lack of surface water. Pollution indi-

TABLE 7.4
Many urban river sections do not even meet grade 5 standards, 1995
(percent)

Region	Grade 2	Grade 3	Grade 4	Grade 5	Below grade 5
North (66 observations)	4.6	3.0	18.2	22.7	51.5
South (69 observations)	17.4	18.8	29.0	8.7	26.1
National (135 observations)	11.1	11.1	23.7	15.6	38.5

Source: NEPA 1997.

TABLE 7.5
Urban river sections contain a variety of pollutants, 1995

Pollutant	Grade 3 standard (milligrams per liter)	Urban river sections with annual average concentrations exceeding grade 3 standard (percent)
Petroleum	0.05	70.5
Permanganate	6.0	51.1
Biological oxygen demand	4.0	44.7
Ammonia nitrogen	1.0	44.4
Volatile phenols	0.005	37.3
Dissolved oxygen	5.0	31.9
Mercury	0.0001	18.8
Cadmium	0.005	7.7
Lead	0.05	7.0

Source: NEPA 1997.

cators such as total hardness, degree of mineralization, and nitrate concentration have increased. Pollutants such as volatile phenols, cyanide, and nitrate are found in groundwater, indicating the seepage of polluted surface water.

Coastal waters have seen a rapid decline in quality. Inorganic nitrogen pollution is serious and widespread, with half of 839 monitored samples exceeding the maximum limits (NEPA 1996).[3] Major estuaries, bays, harbors, and waters near medium-size and large cities are the most polluted. For example, in the Yangtze River Estuary, Hangzhou Bay, and Zhoushan fishing ground areas inorganic nitrogen exceeds the maximum limit by 100 percent and phosphorus, by 87 percent (NEPA 1996).

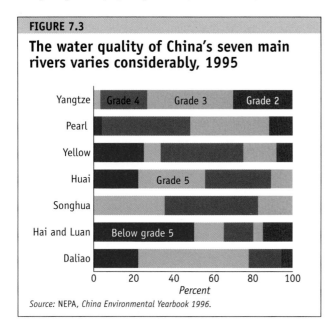

FIGURE 7.3
The water quality of China's seven main rivers varies considerably, 1995

Source: NEPA, *China Environmental Yearbook 1996.*

In summary, the waters in China, with few exceptions, have seen much of their assimilative capacity exhausted by growing pollution. Water quality in major water systems has deteriorated since the early 1980s, with a significant portion falling below grade 3 (maps 5 and 6).

Because China has managed to provide safe drinking water to its large population, it has been able to avoid an epidemic of waterborne diseases. But unless pollution control efforts are stepped up, the quality of China's rivers, lakes, groundwater, and coastal waters will continue to deteriorate—with potentially serious long-term effects.

The government recognizes the potentially dire consequence of inaction. Regulation has improved markedly since the late 1980s and in recent years has stabilized pollution levels in some major rivers. A large-scale cleanup effort for the Huai River Basin started in 1995, and similar programs are being implemented or planned for other heavily polluted major rivers and lakes. Through these and similar efforts, China hopes to reverse the trend of deteriorating water quality by early in the next century.

Water resource management and pollution control regulation

Both national and local institutions are responsible for water resources and pollution control management. The Water Law and the Water Pollution Prevention and Control Law assign these responsibilities and provide the legislative underpinnings of China's water management regulations.

National agencies

The Ministry of Water Resources is responsible for national water resources management, including improving and developing major rivers, planning water resources for urban water supplies, constructing basic rural irrigation facilities, implementing soil and water conservation programs, supplying rural hydropower, and building and managing medium-size and large reservoirs for flood control, irrigation, water supply, and rural hydropower. The ministry oversees seven river basin commissions that coordinate regional water resource management activities. Provincial, municipal, and county governments maintain relatively autonomous water resource bureaus.

The National Environmental Protection Agency is charged with developing water pollution regulations to protect national water quality. These regulations are enforced primarily by local environmental protection bureaus.

The Ministry of Construction and its local counterparts play a key role in planning and constructing municipal water systems, including sewage collection and treatment facilities.

Because of its large hydropower operations, the Ministry of Electric Power is also involved in managing water resources. As with the other ministries, it also draws on its regional and local agencies.

A few other national agencies and their local counterparts are also involved with managing, developing, or protecting water resources. The Ministry of Transportation develops and manages inland rivers and coastal waters for navigational activities. The Ministry of Geology explores groundwater resources. The Ministry of Public Health monitors drinking water quality and waterborne diseases.

The National Coordination Group on Water Resources was established to unify the management of water resources, enhance information exchange, and facilitate coordination among agencies. The group is headed by the minister of water resources and has as members the ministers of all the concerned ministries, as well as representatives from the National Environmental Protection Agency, the State Planning Commission, and the Academy of Sciences. Several provinces and municipalities have established similar water resource committees.

River basin management

The river basin commission is the principal administrative, advisory, and consulting agency within a river basin. The commissions' focus is on flood, sediment, and drought control. In water-short basins the commissions often must resolve interprovincial water conflicts within the basin and monitor water quality in coordination with relevant ministries and environmental protection agencies. The commissions tend to be well staffed. For example, the Yellow River Conservancy Commission employs about 29,000 people throughout the basin.

Provincial water resource bureaus are responsible for planning, surveying, designing, constructing, operating, and managing irrigation, drainage, flood control works, and rural hydropower. Bureaus at the municipal and county levels are responsible for constructing and maintaining most canals, associated irrigation and flood control structures, and medium-size reservoirs. Townships and villages share responsibility for constructing and maintaining branch canals, ancillary works, and small reservoirs. Staff of municipal and county water resource bureaus are organized into irrigation districts responsible for operating and maintaining projects, planning and distributing water, and levying water charges. At the township and village level water users elect members to canal committees. These committees handle on-farm water management under the direction of irrigation district staff.

Water pollution control is essentially undertaken by the provinces through municipal and county governments. The environmental protection bureaus at each level are responsible for implementing national and local water pollution regulations. Water pollution control along the borders of a river basin has not been dealt with effectively, however. Boundary-area water quality is monitored by the water protection institutes of river basin commissions. But the institutes are often underfunded and do not have enforcement authority. The interim arrangements for Huai River Basin are the first attempt to control water pollution on a basinwide level (box 7.1).

Urban water supply and wastewater treatment

The Ministry of Construction regulates investment in municipal water supply and wastewater treatment; major investments require ministry approval. Municipal water supply responsibilities are usually shared by city governments and enterprises. Enterprise self-extraction of water, both for production and for employees, accounted for about half of the urban water supply in the early 1990s.

Urban water supply has long been considered a basic government service. Water quotas have been the norm for demand management. Prices traditionally have not reflected resource value and cost recovery. Self-extracted water charges are even lower (World Bank 1994). Reforms have been introduced to raise water prices and self-extraction fees and reduce the reliance on quotas. But subsidized water supply, even for commercial users, is still common in Chinese cities. Cheap water has led to

Interim water pollution control regulation for the Huai River Basin

The Huai River Basin water pollution control program was initiated in 1995. The Ministry of Water Resources and the National Environmental Protection Agency head the program's central government leading group, which includes members from the four basin provinces and other relevant central agencies. A basinwide pollution control plan has been formulated and interim regulations promulgated. The government considers the Huai initiative a model for water pollution control in other river basins. The initiative represents a major step forward in China's water pollution control policy:

• For the first time pollution control is being undertaken on a basinwide level and is being jointly implemented by the National Environmental Protection Agency and the Ministry of Water Resources.

• Pollution loads are defined by province, county, and township for sections of the river basin.

• Key polluting industries are targeted for reduction of pollution load, and heavily polluting and unprofitable enterprises will be closed down.

• Time limits are provided for polluting industries to comply.

• Local governments will incorporate the pollution load plan into their economic development plans.

• All charges collected from pollution fines and other charges will be channeled directly to pollution abatement facilities.

• A monitoring system will be established for industries to report their pollution loads.

• Water quality monitoring along borders will be done by the Huai River Basin Commission, with results reported to the leading group.

Although the initiative is the first step in a major effort to control water pollution, as things stand the plan may not be sustainable:

• The costs and financing of load reductions are not dealt with in the interim plan. The cost of shutting down industries will have to be borne by local governments at different levels. These costs can be enormous, and financial arrangements will have to be made between the central and provincial governments to finance some of these activities.

• The interim plan does not say how industries are going to reduce their loads. The technical expertise does not exist, especially at the county and township levels, for industrial pollution control or prevention. Technical assistance should be provided to each county and township to help them come up with meaningful pollution load reduction plans.

• There is no plan to unify municipal and industrial waste treatment so as to reduce the unit cost of treatment and make it more affordable to industries. Cities and townships should ensure that all industrial wastewater, after inplant pretreatment to remove toxic substances and heavy metals, flows to a municipal sewerage system. The plan also does not discuss sewerage standards for industrial waste.

• There is no plan to unify the city agencies that collect and treat sewage. Collection systems are part of the drainage department of the municipalities, funded in part by the districts of the municipalities. Treatment plants are run by separate departments.

• River water quality monitoring is fragmented into agencies at county, municipal, and provincial environmental protection bureaus and the river basin agency for boundary sections. Such fragmentation can create a lot of confusion because monitoring quality and standards can vary considerably between different levels of government.

Source: World Bank staff.

widespread inefficiency in water utilization, encouraged industries to adopt water-intensive technologies, and generated inadequate funds for water investment.

In the late 1980s 57 percent of municipal sewage was collected. The situation is unlikely to have improved much given the explosive growth of cities and towns in the 1990s. Municipal wastewater treatment is even lower because of a lack of investment funds. In 1992 only 4.5 percent of municipal wastewater received any kind of treatment. Including treated industrial wastewater, the amount of treated urban wastewater rises to about 17 percent (World Bank 1994).

The National Environmental Protection Agency and local environmental protection bureaus regulate and enforce industrial water pollution control, which has mainly been targeted at state-owned enterprises, often the main sources of local pollution.[4] The rapid growth of township and village industrial enterprises has spread what used to be a relatively concentrated urban problem to a much wider area. Many of the small tributaries of main rivers are now heavily polluted by effluents from these enterprises, which are largely unregulated.

Problems with the current framework

Although the water management system is elaborate, a number of weaknesses must be addressed.

First, there is no legal framework for managing water resources along river basin boundaries. The Ministry of Water Resources has drafted a directive to address this shortcoming. This draft regulation should be promulgated as soon as possible.

Second, except for flood control, there is no real-time management of water resources in the seven major river basins. Real-time management must be implemented for catchments and subcatchments, especially in provinces where water is in short supply.

Third, there is little agency coordination on water projects in the seven major river basins.[5]

Fourth, within provinces water resource planning is designed to maximize provincial benefits. In some cases this results in suboptimal basin benefits. River basin commissions should have more authority in examining the provincial plans and fitting the plans into the overall basin plan.

Fifth, water resource charges are extremely low. The river basin commissions cannot meet their operating costs from these low charges. The commissions are not allowed to charge for regulating dams or for flood control or sediment control. Large sums of central government money are used to subsidize the commissions' operations, and this is not financially sustainable.

Finally, the water pollution control responsibilities of the river basin commissions only cover boundary areas. The commissions' water protection institutes have very little funding to monitor pollution even in the river sections that they are authorized to monitor. The commissions should monitor all parts of the river basin, including the tributaries. They also should collect and collate water quality data from provincial, municipal, and county environmental protection bureaus to obtain a comprehensive understanding of river basin water quality.

The separation of water resource management and pollution control and the fragmentation of national and local agencies are the main institutional problems of the water sector. The government is adopting policies to ameliorate these problems.

Water resource management and pollution control in other countries

Most industrial countries have addressed water resource management and pollution control at the same time. Several lessons can be learned from these countries' experience:

• Water resource management and water pollution control should be vested in river basin authorities, which should be corporatized and should have the authority for water abstraction, flood control, drainage, effluent discharge into any part of the river system, water quality maintenance (which requires a basinwide water quality monitoring system), navigation, coastal pollution, and water flow management. These entities should also be responsible for collecting water abstracting charges (similar to China's water resource fee), including monitoring charges; flood control and drainage charges; and effluent charges and fines for discharges into river waters.

• All cities and towns should have an integrated sewerage system for municipal and industrial wastewater. Industries should have inplant treatment to remove toxic materials and heavy metals before discharging into the municipal sewer system. No discharge to open watercourses from industry should be allowed.

• Charges for water abstraction, water supply, sewage treatment, flood control, and drainage should be set high enough to recover the full costs of these operations.

Costs and benefits of a clean and adequate water supply

Limited water resources can be managed in ways that ensure a safe and sufficient supply without imposing enormous costs on the government. Two mechanisms that can achieve this goal are higher water prices and wastewater treatment efforts.

Raising water prices

Many cities in northern China are using pricing and technical assistance programs to reduce water shortages, manage demand, and reduce losses in water systems. Although water prices have been low for all users, provincial and state governments are encouraging price increases that reflect both cost recovery and shortage costs. In Taiyuan (Shanxi Province) the Price Bureau has announced that water prices will increase from an average of 1.24 yuan a cubic meter to 4.94 yuan a cubic meter over the next five years in order to recover supply costs. Many cities in Hebei Province have proposed an increase in the water tariff to recover supply costs. Water tariffs in these cities range from 0.5–0.9 yuan a cubic meter (table 7.6), while the marginal cost of new water projects is generally above 1.20 yuan a cubic meter.

TABLE 7.6
Water tariffs are low in selected cities of Hebei Province, 1996
(yuan per cubic meter)

User	Handan	Hengshui	Langfang	Qinhuangdao	Shijiazhuang	Tangshan
Residential	0.50	0.45	0.40	0.40	0.65	0.50
Government	0.50	0.75	0.80	0.60	0.60	0.50
Industrial[a]	0.80	1.00	1.00	0.80	1.20	0.70
Commercial	1.25	—	1.50	2.50	1.40	1.20
Average	0.63	0.85	0.76	0.73	0.90	0.56

a. The charge for self-extracted water is much lower.
Source: World Bank staff estimates.

The willingness to pay for urban water is estimated to be about 1.8 yuan a cubic meter in some coastal cities and about 4.3 yuan a cubic meter in the extremely water-short city of Taiyuan, which is representative of water-short cities in the Hai River Basin. The economic value of water across industries in Shanxi and throughout the Hai River Basin is about 24 yuan a cubic meter. Water shortages in Shanxi, China's coal mining base, are so severe that they threaten the growth of local industries (box 7.2).

Increasing wastewater treatment

Many urban areas will face increasing shortages of grades 1, 2, and 3 water if the quality and quantity of municipal and industrial wastewater treatment do not rise. Safe drinking water extraction is already a serious problem along the major rivers in northern China. Large water transfer projects require huge investments and take years to complete. In cities and towns where water shortages are caused primarily by water pollution, wastewater treatment provides basic health benefits and contributes to industrial and agricultural productivity. Increased water supply and water consumption will only lead to greater wastewater emissions if effluent treatment does not catch up.

World Bank studies suggest that primary treatment (which typically removes about 30 percent of biological oxygen demand and most heavy metals in wastewater) would require a tariff of 0.50 yuan per cubic meter of water (1994 yuan) to recover costs. Adding secondary treatment (which removes 85–95 percent of biological oxygen demand) boosts the tariff to 0.77 yuan per cubic meter. Imposing this treatment charge on users would increase the cost of using water by 50 percent or more but could be easily absorbed by most users (World Bank

1994). Raising wastewater treatment prices toward full cost recovery would generate badly needed funds for expanding urban wastewater treatment facilities. In water-short cities wastewater treatment would provide a cheaper water supply alternative to some industrial users (box 7.3).

Since irrigated agriculture is the main user of polluted water, uncontrolled wastewater discharges into rivers and lakes have significant consequences for agriculture. In general, agriculture cannot use irrigation water below grade 5 standards because it contains large amounts of heavy metals and toxins. In 1993 about 8 percent of China's irrigated areas received water below grade 5 standards, resulting in an estimated 1 million

---

BOX 7.2

Industries are conserving water in Shanxi Province, but supply is still a problem

Shanxi industries have gone to great lengths to conserve water. In 1980 the average output value per cubic meter of water was 22 yuan. By 1990 this had risen to 48 yuan, and by 1994 it reached 75 yuan. Shanxi industries surpassed nationwide recycling goals for 2000 in 1991, achieving an 84 percent recycling rate in 1993. In recent years most investment in Shanxi industries went toward increasing the productivity of water. Still, water remains the overriding constraint on industry. It is estimated that industrial output in 1993 could have been 6.5 billion yuan greater (22 percent more) had there been no water shortage. Factories periodically have to shut down when water is not available.

If water shortages are alleviated, Shanxi's industrial capacity will continue to grow based on its location and endowment of natural resources. With additional water, annual industrial output is projected to increase by 75 percent between 1993 and 2000. Without additional water, output is likely to stagnate and remain near current levels.

Source: World Bank staff.

tons of lost grain production (NEPA 1997). Since irrigation peaks during the dry season, when natural dilution capacity is at a minimum, industrial and municipal wastewater generally should be treated up to grade 5 quality for irrigation purposes, especially in northern regions. Current treatment fails to meet this requirement by a substantial margin, with the chemical oxygen demand concentration of industrial discharges exceeding the grade 5 standard by thirteen times. Lost grain production could increase significantly if treatment of industrial and municipal wastewater does not improve, affecting China's food supply.

The effect of wastewater treatment on China's future grain production has been evaluated with a water demand model covering China's nine major river basins. Treated wastewater is assumed to account for about 30 percent of the total effluent volume by 2000, from less than 20 percent today. Thus the model allowed for projected treatment levels to vary from 30 to 75 percent. Net grain production under different treatment levels are shown in figure 7.4. Increasing wastewater treatment from 30 to 50 percent would increase grain production by about 24 million tons by 2020—enough to support 65 million people (at 1996 per capita gross grain consumption levels). The additional gain is much smaller if the treatment level is further increased from 50 to 75 percent—about 7 million tons.

Policies for sustainable use of water resources

China faces enormous challenges in resolving chronic water shortages in the north and controlling widespread water pollution. Policies to speed up investment in water supply and wastewater treatment are a top priority. Prices for freshwater supply and charges for effluent treatment are increasing but much more needs to be done to achieve an economically sound and socially efficient tariff structure. National and local water institutions are the key to successful water policies. A basin-oriented management system with strong central and regional leadership and clearly defined local responsibilities will be essential for integrated water resource and pollution control management.

Managing demand by raising tariffs and treatment fees

Despite recent water pricing and management reforms, China's water supply and wastewater treatment services are generally underpriced, leading to excess demand, high pollution, and inadequate funds to meet investment needs. International experience has shown that water users are sensitive to water prices. Among households the price elasticity of water demand varies between –0.3 and –0.6, meaning that a 1 percent increase in water prices reduces demand by 0.3 to 0.6 percent. In the industrial sector the range is –0.45 to

---

BOX 7.3

Treating water is about half as costly as securing new freshwater in Dalian

Dalian, a coastal city, suffers from water shortages. About 40–50 percent of the city's tapwater is consumed by industry. During the dry season the government used to order some large industrial users to cease operations for as long as sixty days.

In 1992 a wastewater treatment plant with treatment capacity of 100,000 cubic meters a day was put into operation, supplying processing water for one of the city's industrial districts. The treated wastewater (meeting grade 4 standards) provides factories with a low-priced alternative water source and has boosted production and reduced pollution.

Best of all, treatment is affordable. Dalian's industries pay 1.5 yuan per ton of tapwater, which has a supply cost of 1.2 yuan per ton. Treated wastewater, by contrast, is priced at 0.8 yuan per ton (treatment costs 0.5 yuan per ton).

Source: Rui 1997.

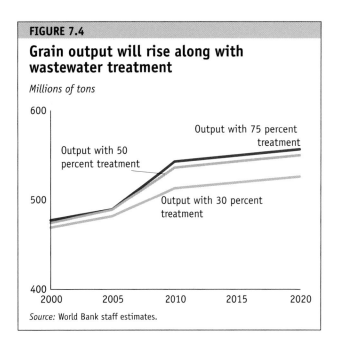

FIGURE 7.4

Grain output will rise along with wastewater treatment

Millions of tons

Output with 75 percent treatment

Output with 50 percent treatment

Output with 30 percent treatment

Source: World Bank staff estimates.

–1.37. And for irrigation the elasticities fall between –0.37 and –1.50 (World Bank 1995). Higher prices would encourage large water consumers in industry and agriculture to adopt more efficient water use practices and technologies.

Water must be priced to reflect its scarcity and economic costs. The Ministry of Water Resources recently recommended freeing water prices, starting with all new water projects. More aggressive local governments have already started to allow fully self-financed water projects to set their own prices. Economic water pricing will enhance demand management and provide the financial resources water companies need to increase maintenance and service quality and expand system coverage.

Groundwater and surface water must be protected from overexploitation. Since half of the urban water supply is used by enterprises, fees for raw water extraction should be increased to a level that equates self-supply costs with municipal supply costs. Water demand management that uses economic pricing of water extraction and supply would reduce investment needs, free enterprises from water quotas, and allow cities to reallocate much of the urban infrastructure budget that currently finances water investments.

To prevent further deterioration in water quality, cities and towns must increase municipal wastewater treatment levels at a pace at least equal to the growth in effluent discharges. Wastewater treatment plants are the only effective means of dealing with the growing domestic and commercial wastewater loads in urban areas. General government revenues will not be nearly enough to meet the required investments in treatment. Cities must raise sewerage fees for domestic, commercial, and enterprise discharges to levels that recover the costs of collection and treatment. While imposing user fees may be politically difficult, city leaders must recognize that the costs ultimately will be paid directly or indirectly, and that relying on indirect payments leads to excess effluent discharges and makes it impossible to finance treatment needs. The result is more pollution and higher social costs.

Controlling discharges at their source is crucial. Enterprises will not have an incentive to reduce pollution unless it becomes an internal cost through effluent levies, violation fines, and enforcement monitoring. Levies and fines for water pollution are universally low in China, and should be raised substantially to induce greater abatement. Controlling water pollution from small industrial enterprises will be difficult. In the short term it will be necessary to close highly polluting small factories. In the long term, however, collective treatment of these enterprises' discharges and municipal or small-town wastewater is a viable solution.

The absence of adequate planning undermines water resources management. Most Chinese cities do not have watershed management master plans that integrate water supply, demand management, and pollution prevention and control. Water supply investment is generally based on a needs approach, without much consideration of the role of water prices and their effect on consumption. China's cities and regions must improve their planning capacities by adopting new methods and taking a comprehensive approach to watershed management.

Improving irrigation efficiency should be the top priority of demand management for agricultural water use, and could be accelerated with government technical assistance programs that disseminate water-saving irrigation technologies.

Ensuring supply by increasing investment

Only by increasing investment can China ensure a reliable water supply and improve water quality. To relieve the severe water shortages in northern China, it may be necessary to invest in south-north water transfer projects. But such projects are no substitute for using water resources as efficiently as possible, through conservation, recycling, and wastewater treatment. All cities will need capital to strengthen water supply and sanitation services. These investments will reduce the economic losses due to water shortages and lower the social costs of water pollution. The government cannot shoulder all of the investment costs. Proper pricing policies and environmental regulation can help realize the potential for private investment in the sector.

A major effort must be made to install sewerage in all cities to ensure that all waste (municipal and industrial) is collected. Treatment plants should also be built. Financing for collection and treatment may have to be obtained by issuing bonds or by attracting private or public investments. Cities and towns in Britain and France initially received loans from the central government for such undertakings; over time the companies generated enough cash to continue the programs. In the end these water and sewerage investments became financially attractive to private companies.

China's central government should ensure that within ten years all medium-size and large cities provide primary treatment (with provisions to expand to secondary) for any discharge that lowers water quality below grade 5 standards. All cities should be required to develop sewerage systems that facilitate the later addition of treatment works and that provide immediate treatment if water does not meet standards for irrigation within 3 kilometers of wastewater discharge points or standards for domestic raw water at the nearest downstream community extraction point. The costs of investing in urban wastewater collection and treatment will be offset by a decrease in water supply investment resulting from better water demand management (World Bank 1994).

Reforming institutions by integrating resource management and pollution control

Institutional reform should aim to facilitate the integration of water resource management and pollution control on a basinwide basis. This in turn will require stronger institutions for basin management.

River basin commissions. The seven river basin commissions should be corporatized and become legally responsible for the following functions in the basin:
- Water abstraction.
- Flood control.
- Drainage.
- Navigation.
- Hydropower discharges.
- Water flow maintenance.
- Water quality maintenance of all main and tributary rivers.
- Coastal pollution.

The commissions should be able to set water tariffs for each function so that they can recover their operational costs and develop new projects to control water flows in the river basin. A portion of the charges for effluent control and water abstraction should be channeled to cities and industry for water pollution control.

The river basin commissions should establish a water quality monitoring system for the entire basin (including the tributaries) for daily, weekly, and monthly monitoring. For regulatory purposes, the National Environmental Protection Agency and local environmental protection bureaus should set up an independent primary network for monthly or random monitoring to verify the measurements undertaken by the river basin commissions.

Local governments. Cities and towns should integrate wastewater collection and treatment systems for municipal and industrial waste. In the long term discharges by any point-source polluter into any open body of water should not be permitted. All point-source discharges should be collected through municipal sewerage and treated. The unit cost of treatment for industries using an integrated system will drop to 1–5 percent of the cost of self-treatment because of the system's huge economies of scale. Pretreatment for industrial wastewater should be required to remove toxins and heavy metals before the water is discharged into sewers. Sewer discharge standards should be developed and implemented. Wastewater charges for industry discharging into sewers should be based on volumetric charges for primary and secondary treatment, chemical oxygen demand and biological oxygen demand levels and loads, and suspended solids and the cost of sludge disposal.

Wastewater collection and treatment should be the responsibility of one agency in the city government. This sewerage agency should be corporatized and made financially autonomous. Sewerage charges should be included in water bills and transferred to the sewerage agency.

The National Environmental Protection Agency and local environmental protection bureaus. The roles of the National Environmental Protection Agency and the environmental protection bureaus at different levels of government should be expanded to regulate and enforce all environmental standards. Both entities should:
- Elaborate and supplement the regulatory framework through legislation that establishes standards for water supply and sewerage services, sets out criteria for classifying river water quality, and imposes requirements to protect against pollution.

- Approve codes in various aspects of conservation, recreation, ecology, and so on.
- Ensure that water and sewerage companies undertake work to the standards required and, if these standards are continuously violated, prosecute any water and sewerage company.
- Appoint technical assessors of water quality in river or coastal areas.
- Oversee waste disposal authorities, especially for toxic waste.
- Provide independent advice at different levels on pollution control practices.
- Approve, inspect, and oversee potentially polluting processes and those that endanger the safety of workers.

Ministry of Water Resources. The Ministry of Water Resources's role should be expanded to:
- Oversee river basin developments by the basin authorities.
- Approve all water abstraction charges.
- Prepare water dispatching systems for all states of river water (flood, ice flood, drought, sediment) and have complete authority over all river regulation (including discharges from major hydropower stations).
- Develop water protection zones and ensure that these zones receive minimum supplies.

Ministry of Construction. The Ministry of Construction's role should be expanded to:
- Carry out technical audits of water and sewerage companies in cities.
- Investigate incidents that adversely affect drinking water.
- Provide technical and scientific advice on drinking water policy.
- Identify hazards relating to drinking water and initiate research.
- Provide guidance on analytical methods used in monitoring drinking water.
- Review all major plans for water supplies and assess the chemicals and materials used in connection with these supplies.

A role for the public. Water management and pollution control are public services that require public support and public participation. Public involvement in the legislative and regulatory processes affecting water issues will improve the efficiency and effectiveness of water management.

Notes

1. Water resources can be divided into surface water and groundwater. But except for deep deposits (sometimes called fossil water), groundwater occurs only as a result of seepage of surface water runoff.

2. These number add up to more than 270 because some cities suffer from multiple problems.

3. The National Environmental Protection Agency classifies three grades of seawater quality. Marine nature reserves must obtain grade 1 quality, grade 2 seawater permits swimming, and grade 3 permits industrial uses.

4. The National Environmental Protection Agency regulates wastewater discharges through three classes of discharge standards. Class 1 standards apply to discharges into key protected water bodies, including drinking water sources, general fishery areas, and important scenic areas. Class 2 standards apply to discharges into generally protected water bodies, including general industrial water sources and irrigation water sources. Class 3 standards apply to discharges receiving secondary wastewater treatment. Despite the reported increase in treatment, chemical oxygen demand concentrations in wastewater discharged by regulated enterprises rose from 250 milligrams a liter in 1986 to about 350 milligrams a liter in 1995, compared with the class 2 standard of 100 milligrams a liter. Wastewater discharged by township and village industrial enterprises is believed to be significantly more polluted because of the near-absence of treatment.

5. Projects on the main stem of the river basin are operated by individual provinces or the relevant ministries. The operation of a major power dam follows the load dispatch requirements of the regional electric grid. Although some arrangements have been made to ensure that flows are maintained in certain months of the year, these cannot be changed. There are no real-time operating rules, which are under the control of the river basin commissions. For example, the Yellow River Conservancy Commission has an explicit water allocation for each province, but there is no control over the abstraction rates by the provinces. Hence the Yellow River has been drying up for as long as 130 days of the year for as much as 600 kilometers of the lower-reach sections. The basin commission's power has been limited to flood and sediment control for the lower reach of the basin. It is not even authorized to develop a water dispatching system for the whole basin.

References

Chen, Xiaojun. 1996. Personal communication. Senior Economist, Ministry of Water Resources, Beijing.

Lu, Dadao. 1993. "An Important Solution to the Problem of Water Shortage in North China." *Chinese Environment and Development* 4(3).

MOT (Ministry of Transportation). 1997. "Mobile Pollution Sources on Chinese Rivers." Background paper submitted to the World Bank. Beijing.

NEPA (National Environmental Protection Agency). 1996. *Report on the State of the Environment of China 1995.* Beijing.

———.1997. "China's Water Quantity and Quality Issues." Background paper submitted to the World Bank. Beijing.

———. Various years. *China Environmental Yearbook.* Beijing.

Rui, Haifeng. 1997. Personal communication. China Academy of Sciences, Beijing.

SPC (State Planning Commission). 1995. *Report on China's Population, Resources and Environment.* Beijing: China Environmental Sciences Press.

SSB (State Statistical Bureau). 1996. *China Statistical Yearbook 1996.* Beijing.

World Bank. 1993. "China Yellow River Basin Investment Planning Study." Report 11146-CHA. East Asia and Pacific Regional Office, Washington, DC.

———. 1994. "China Urban Environmental Service Management." Report 13073-CHA. East Asia and Pacific Regional Office, Washington, D.C.

———. 1995. "Water Conservation and Reallocation: Best Practice Cases in Improving Economic Efficiency and Environmental Quality." In *Water and Sanitation Currents.* UNDP–World Bank Water and Sanitation Program. Washington, D.C.

———.1997. "China Wanjiazhai Water Transfer Project." Report 15999-CHA. Washington, D.C.

WRI (World Resources Institute). 1996. *World Resources, 1996–97.* Washington, D.C.

Wu, Yiao, ed. 1989. *Water Resources Utilization in China.* Beijing: Ministry of Water Utilization and Power Generation.

chapter eight

Policies: Securing Higher Environmental Living Standards

The spread of markets in China has driven economic growth, leading to higher incomes, lower poverty, and improvements in nearly all aspects of Chinese life. But growth has also increased the pressures on China's environment.

Awareness of these problems triggered new environmental protection efforts starting in the late 1970s. Trickles of policy change have since become a steady stream of new initiatives, policies, and regulations. China has enacted more than a dozen significant pieces of legislation to safeguard the environment, and each year the government has invested more resources to protect the environment. These efforts, together with economic reforms, have reduced the pollution intensity of Chinese production. Air and water quality, though still far below Chinese aspirations, have been held in check despite the unrelenting pressures of an expanding economy.

But as Premier Li Peng noted in his speech to the Fourth National Environmental Conference in 1996, much remains to be done. Air and water pollution levels in China are among the highest in the world and, according to World Bank estimates, cause $54 billion a year—about 8 percent of GDP—in damages to human health and lost agricultural productivity.[1] Yet the same rapid growth that has added to China's environmental woes in the past provides the opportunity to boost environmental living standards in the future. With a bit more investment (see chapter 3), China can "grow greener." What should be the priorities, and what policies will elicit the required investments?

Establishing priorities for policy

Estimates of pollution-related damages to health, worker productivity, and agriculture (presented in chapter 2 and projected as future scenarios in chapter 3) point to several environmental priorities:

• Among the various types of pollution, urban air pollution causes the most damage—$33 billion a year, mostly through its effects on human health. Without assertive policies, future emission reductions from large point sources, such as power plants, could be offset by worsening emissions from small point sources. Increasing automobile use will also have an adverse effect on urban environmental quality.

• Severe indoor air pollution from burning coal and biomass causes about $10 billion a year in health and mortality damages.

• The costs of water pollution, difficult to quantify and probably underestimated in this report, are at least $4 billion a year. These costs will grow as water shortages become more acute, municipal waste increases with urbanization, and agriculture practices (particularly fertilizer and pesticide use) continue to generate runoff pollution.

• Acid rain, responsible for $5 billion a year in damages, is also likely to worsen because of China's dependence on coal.

• The government's decision to phase out lead from gasoline is important. Now it must be swiftly implemented to prevent irreversible damage to the intelligence and neurological development of Chinese children and save costs of neonatal care, medical treatment, compensatory education, and lower productivity.

Addressing these challenges will require a new strategy for environmental protection that reflects China's increased market orientation, dynamic growth patterns, and extensive administrative capacity. These new policies must:

• Harness the market to work for the environment, not against it. This will require adjusting prices to cover economic costs and incorporating social costs through pollution taxes.

• Harness growth for the environment. This will require creating incentives to elicit investments—from both the commercial state and nonstate sectors and from the public sector—with the largest environmental benefits for future generations.

• Harness the government's administrative capabilities to the cause of the environment. This will require providing better regulations but with fewer investment controls at the national level as well as ensuring better planning and pollution enforcement in urban airsheds and regional water basins.

Harnessing the market: Making polluters pay

Unregulated market forces are not much better than a command economy in protecting the environment. In its legislation the government has recognized that polluters, rather than victims, should pay for pollution. Yet the full potential of this principle has not been realized.

The prerequisite: Economic reform

Harnessing the market requires, first, accelerating economic reform of state enterprises, the pricing system, and international trade. Government efforts to increase competition (especially from nonstate firms), improve price signals, and expand international trade have made major contributions to lower pollution intensities over the past decade (chapter 5). Yet other, related reforms remain:

• Reforming state enterprises would make them more responsive to market signals, including pollution taxes and regulatory penalties; it is nearly impossible to make polluters pay when enterprises can fund their pollution taxes or levies through implicit subsidies from the treasury or financial system.

- Expanding the realm of competitively determined prices, including the price of investment capital, would channel resources to efficient firms; firms that have privileged, noncompetitive access to capital waste resources at a high cost to the environment.
- Extending trade reform would end pockets of protection for industries and put pressure on inefficient industries, many of which are pollution-intensive.
- Deepening domestic capital markets and establishing a framework for private participation in infrastructure would open up new ways for municipalities to finance their investments in water supply and sanitation. These measures would also facilitate investment in energy-efficient equipment that embodies abatement technology.

Since these reforms touch every aspect of Chinese economic life, they will take time to complete. Nonetheless, accelerating the pace of reform will reduce the cost of cleanup and rapidly improve China's environmental living standards.

Pricing to cover costs

Despite the fact that China is resource-poor, it prices its energy and water like the resource-rich United States. Over the past three years the government has raised and partly deregulated coal prices, and in most areas coal prices now cover economic costs. In many places, however, market prices do not fully reflect the sulfur and ash content of coal. Markets are distorted because the planning system continues to allocate a sizable portion of coal to certain activities and firms; because the coal transport system is overburdened, antiquated, and saddled with anticompetitive regulations; and because many small, unregulated mines, unburdened by the costs of safety and environmental regulations and worker benefits, produce low-quality coal that undercuts the price of quality coal. Pricing reform and reform of the production and distribution system must go hand in hand.

Natural gas remains heavily regulated. This is done to provide supplies to the fertilizer industry at below-market prices. Low prices do not, however, provide incentives for state-owned, much less private, gas suppliers to enter the market. Moreover, households are willing to pay more for this clean heating and cooking fuel. Deregulation and market-based pricing are needed to substitute for household uses of coal.

China also allocates its domestic crude oil below international prices. To develop alternatives to coal, the government should establish a transparent pricing framework for oil and gas and ease entry to all parts of the oil and gas industry for domestic and foreign investors alike.

Pricing to cover the economic costs of production will reinforce current efforts to encourage energy conservation. Efficient pricing will induce more cost-effective conservation in several activities: cogenerating steam and electricity, renovating furnaces and kilns, managing energy systems, and recovering waste heat, combustible gas, and materials from industrial processes.

Water pricing will soon become even more of a major policy issue. Although one of its most precious resources, China wastes a lot of water by failing to charge full value for water supply and wastewater treatment services. China also lacks the frequent monitoring and substantially higher fines and other controls that are needed to enforce industrial wastewater discharge standards. Although not always possible for political and practical reasons, China should continue adjusting sewerage user fees toward full cost recovery. Targeted subsidies can be carefully designed and provided to low-income users.

Taxing environmental externalities

Pollution taxes and charges are another way to harness the market to protect the environment (box 8.1). The government should consider taxing coal in a way that reflects its enormous social costs. If health damages were included in its price, coal in Beijing would cost 100 percent more (see chapter 4). A tax based on sulfur and ash content would deter the use of poor-quality coal and make competing fuels more economic. Though it might be preferable to tax emissions rather than inputs, the enormous number of small polluters makes collection difficult. A quality-based coal tax, levied at mines for large producers and distribution points for small producers, would create incentives for users to use cleaner or washed coal and for producers to improve coal quality. Such a tax could be announced and phased in over time so that current investment decisions—in boilers, space heating, and other coal-consuming equipment—can anticipate future prices and invest in clean technology accordingly.

China does have pollution levies, but they are poorly designed and inadequately enforced. In fact, levies are so low they do not even encourage firms to make cost-effective investments in abatement. As a result the overall abatement rate is far below the optimal level. Data for Zhengzhou indicate that reducing sulfur dioxide emissions by 100 tons costs about $170 and produces benefits (measured in avoided mortality) of about $5,000—an implicit social rate of return to abatement of about 2,800 percent. At such a low abatement cost level, failure to enforce control implicitly values a citizen's life at less than $300 at the margin (see chapter 5).

Still, even today's low levies are encouraging some firms to abate. This responsiveness suggests that by 2020 higher levies could reduce particulate emissions and chemical oxygen demand pollution from regulated industries to less than 20 percent of current levels.

The government's recent proposal for a tenfold increase in the air pollution levy is especially important for controlling emissions from medium-size and large industrial sources. Such a levy would significantly improve air and water quality in China's cities.

Changing the levy system's design also would increase its effectiveness. Today firms are taxed only for the pollutant among their emissions that exceeds the discharge standard by the greatest amount, and then only for the amount over the standard. For both air and water emissions the government should consider modifying the levy to impose charges on all emissions of a few harmful pollutants (such as particulates and sulfur dioxide for air and chemical oxygen demand and biological oxygen demand for water). To increase effectiveness, the central government should give local governments the authority to raise pollution levies to achieve local pollution control goals. Local governments should also be able to decide how to use the levy funds in the context of local conditions. The central government should shift its attention to monitoring and ensuring local compliance with national environmental management requirements. Levies on coal-related air pollution should be coordinated with any externality tax to avoid double taxation.

The power industry is one large point source whose emissions can and should be taxed. A fully enforced air pollution levy would create an incentive for power companies to invest in larger (more than 200 megawatts), more efficient plants. Some 40 percent of the new capacity added to the power grid in the past five years

---

BOX 8.1

Pollution charges: What works and why

Water pollution charge in the Netherlands
The water pollution charge in the Netherlands is determined by how much revenue is required to treat sewage and to maintain and improve general water quality. The charge is imposed by Water Boards—self-governing bodies of surface water users responsible for water management. The charge for industries, based on biological oxygen demand and (in most cases) heavy metal pollution, is levied on all direct and indirect discharges. In 1993 charges across basins ranged from $1.14 to $3.00 per kilogram of discharged biological oxygen demand. Households and small firms pay a fixed amount. Research has found that the pollution charge encourages conservation among large firms that are metered, especially those engaged in livestock production.

Why does the charge work? First, it is rather high, and aims to achieve full cost recovery for sewage treatment. Second, the charge is directly linked to pollution load (for large firms). Finally, the charge program is decentralized and transparent for water users.

Source: World Bank staff.

Sulfur tax in Sweden
Sweden's sulfur tax is levied on diesel fuel and heating oil that contain more than 0.1 percent sulfur, and on coal, coke, and peat. The tax on oil is 27 krona ($4.09) per cubic meter of oil for every 0.1 percent of sulfur content (by weight). For coal and peat the tax is 30 krona ($4.55) per kilogram of sulfur in the fuel. The tax approximates an emissions charge and is refundable if taxpayers can demonstrate a reduction in sulfur emissions. Between 1990 and 1992 the sulfur content of oil decreased nearly 30 percent as a result of the tax. Emissions from burning coal and peat also fell considerably. The tax promoted higher-level flue gas desulfurization, but emissions also fell because users have been substituting for fossil fuels. A tax on carbon dioxide provides an additional incentive to switch away from fossil fuels. Administrative costs are somewhat less than 1 percent of revenue.

Why does the tax work? First, it is high. Moreover, a transparent rebate scheme strengthens the tax's incentive effect. Finally, the program is easy to implement and has low administrative costs. Polluters are responsible for proving their emission levels.

was in small plants (less than 200 megawatts), which have high emissions per unit of electricity. Since these plants will be in operation for at least two more decades, they lock China into a high-pollution growth path. Taxing emissions up to the value of damages would send the right signal for future investment. These taxes would be especially welcome in cities such as Chongqing, where sulfur dioxide emissions are so high they are having measurable effects on surrounding agriculture.

Although China's gasoline prices cover production and import costs, they are low relative to most East Asian and European countries, which use gasoline taxes to pay for roads and reduce environmental damage. China's low gasoline prices damage the environment in two ways. First, automobile-based urban transport consumes more fuel per person-kilometer than alternative travel modes. Second, traffic congestion drives up emissions because road construction cannot keep up with the number of automobiles. Gasoline prices should be set at levels that, in conjunction with other traffic regulations, discourage car use for urban transportation and encourage the use of public transportation. Wider use of less-polluting travel modes would reduce fuel consumption and smog and take pressure off crowded roadways, allowing everyone to travel faster. Similarly, sales taxes and annual licensing fees on automobiles should reflect the cost of the huge public investments in roads and traffic management needed to support automobiles.

Harnessing market forces has two important additional advantages. First, it better allocates financial responsibilities between consumers and commercial producers on the one hand and the government on the other. By using prices, environmental taxes, and levies, the government ensures that polluters and their customers pay for environmental maintenance. If the government were to raise the levy on a redesigned tax base by tenfold and enforce it, tax revenues could reach 37 billion yuan a year ($4.5 billion), or 0.6 percent of GDP in 1995. Second, market-based instruments allow the government to concentrate on monitoring and enforcement rather than on deciding the best abatement investment for a given plant or industry. Thus the government's administrative resources are focused on tax policy and enforcement rather than on the investment decisions of businesses.

Harnessing growth: Investments for a brighter future

Economic growth also should be harnessed to work for the environment, primarily through the cleaner investments that better pricing brings about.

Correct pricing creates incentives for nonstate and commercial state firms to invest in more energy-efficient technology and abatement. China's industrial boilers are only 80 percent as efficient as those in OECD countries, its steelmaking consumes twice as much energy per unit of output as in OECD countries, its fans and pumps are only 85 percent as efficient as advanced international equipment, and its electric motors consume 5–10 percent more electricity to generate the same power (chapter 4). Raising China's industry to international standards will create the opportunity for environmentally friendly investment.

Public investments are also important for harnessing growth for the environment. Small point-source emissions impose the highest costs on human health. About two-thirds of China's urban population still relies on coal for cooking. Gas fuels are two to five times more energy efficient than coal and can drastically reduce emissions. Thus rapid gasification of household fuel consumption is a high-priority environmental policy. Residential consumers prefer gas, so investment in urban gas supply could be quite profitable. This is an activity where public investments could be supplemented with private and even foreign resources.

Finally, to combat rural indoor pollution, the government should invest in education campaigns, increase access to efficient stoves, and supply cleaner coal and coal substitutes for cooking and heating.

Several investments require government coordination; others require the government to absorb risks that private investors will not accept. A top priority is developing a state-of-the-art, clean coal–burning industrial boiler. Since current technology has not yet been adapted to Chinese materials and scales, the government should play an active role in supporting the boiler's development. Other investments are riskier but might have high payoffs. Some renewable energy technologies fall into this category. For example, photovoltaics have the potential to become cost-effective, superior substitutes for coal. Because they are risky, the private sector rarely invests in renewable energy sources, even if prices internalize all environmental externalities. Thus the government should consider investing in pilot projects to ensure experimentation and adaptation to Chinese circumstances. Renewable energy sources may become important—and eventually profitable—as global concerns about greenhouse gas emissions heat up.

In transportation, the government must elicit investments in public transit to ease congestion and smog. The government does not need to actually finance and make the investments, but it must create a regulatory framework that ensures nonstate investments are made. Prices must incorporate externalities (through automobile and gasoline taxes) and be accompanied by expanded public transit alternatives to cars. Otherwise the pricing measures will simply shackle Chinese consumers with the worst of both worlds—traffic jams and high costs. By 2010 the costs of a public transit–based system are projected to be 17 percent less than an automobile-based system—not even counting environmental benefits.

Public investments in water will be among China's most important and most costly. Investing in pollution abatement—whether for industrial or municipal waste treatment—is much cheaper than investing in huge water transfer projects. Even so, projections suggest that even if China creates incentives for firms to abate cost-effectively, water shortages will be so severe that it will have to invest in large-scale south-north water transfer projects (chapter 7).

Given the rapidly increasing pollution effects of municipal wastewater discharges and the near-absence of treatment, Chinese cities should move fast to invest in wastewater collection and treatment facilities. Pricing wastewater discharge and treatment to allow for full cost recovery would attract private investment and free public funds for other urban investment needs.

The investments needed to achieve major improvements in environmental quality are not enormous. Increasing the amount of pollution control investment from the 0.5 percent of GDP assumed in the business as usual scenario to about 1 percent of GDP (the medium-cost investment scenario) would lower projected health-related pollution costs from 13 percent of GDP to 4 percent in 2020. About 40 percent of these investments, mainly in power and industry, would protect airsheds. Investments in water, some 60 percent of the total, would mostly be for municipal wastewater and industrial discharge abatement. Investments of 2 per-

cent of GDP through 2020 would reduce air and water damages to nearly 2 percent of GDP, an environmental standard of living comparable to that in the United States today.

These investments have an economic rate of return that surpasses any conceivable threshold for public investment. Investing 100 yuan in pollution abatement under the medium-cost investment scenario saves 300 yuan in avoided pollution damages relative to the business as usual scenario (see chapter 3). If the effects of future income growth are considered, the return on pollution control investments is even higher.

One of the advantages of shifting to a market-based environmental strategy is that the government would not bear the full cost of these investments. Under the business as usual scenario the government (mainly municipalities) would shoulder 84 percent of investment costs. The rest would come from the nonstate sector and, to a lesser extent, the state commercial sector (for example, state industrial enterprises). But with reforms that widen the scope for nonstate investment in infrastructure and policies that encourage investment in industrial pollution abatement, as in the medium-cost investment scenario, the government's share of environmental investment would be about 66 percent. Under the high investment scenario the government's share would be 59 percent (table 8.1).

Harnessing administrative capabilities: Improving oversight, enforcing standards

China's administrative and organizational capabilities can be used to improve environmental living standards. Even if markets are harnessed for the environment, and even if the government elicits a steady stream of environmentally friendly private and public investments, the government still must regulate and plan with an eye to the environmental future. At the national level this often means fewer investment controls but better regulations and standards. At the local level it means better urban planning and better enforcement of regulations and standards.

Environmental management at all levels of government should be strengthened. This will require coordinating decisionmaking at the national level to implement, monitor, and ensure policy consistency

TABLE 8.1

Environmental investments have enormous returns, 1995–2000

Investment scenario	
Business as usual	
Cost (percentage of GDP)	0.5
State share (percent)	84
Medium-cost investment	
Cost (percentage of GDP)	1.0
State share (percent)	66
Economic return[a]	3 to 1
High investment	
Cost (percentage of GDP)	2.0
State share (percent)	59

a. Present value of accumulated benefits (damage avoided) relative to present value of annualized pollution control costs.
Source: World Bank staff estimates.

among different levels of government. It will require committing more tax resources at all levels—in strategy, policymaking, and scientific research at the national level and in monitoring and enforcement at the regional and local levels. All levels must help disseminate best practices and environmental education and publicize information about environmental quality.

National policies, regulations, and standards

One of the government's most important recent decisions is to phase out leaded gasoline by 2000. In Shanghai, for example, more than half the children are estimated to suffer from blood-lead levels in excess of 10 micrograms per deciliter, the definition of blood poisoning. It takes about ten years to completely eliminate lead from the fuel supply system. Thus rapid implementation must be a top priority.

Another priority is incorporating township and village industrial enterprises into the regulatory framework. Most of these enterprises are ignored in environmental regulation, and in certain industries—coking, small metallurgy, coal mining—they have become a major source of pollution. Pollution could be better controlled if local zoning restricted these enterprises to rural industrial parks where monitoring and control are feasible.

The government could also set standards that incorporate pollution abatement and energy savings. At present cars operating in Beijing are about ten times as polluting as cars in the United States. Tighter standards that push the automobile industry toward the technolog-

ical frontier of cleanliness are essential for reducing these emissions. Similarly, prohibitions on two-stroke motorcycles, major sources of pollution in southern cities, are a priority. Energy efficiency standards should be set for products that are likely to become major drains on energy, such as automobile engines, air conditioners, refrigerators, and other major consumer appliances.

Standards introduced today can markedly lower future energy consumption in residential and commercial buildings. Given the massive building program planned over the next twenty-five years, China has a unique opportunity to improve the energy efficiency of its buildings. Priority should be given to efficient lighting, improved centralized heating, and implementation and enforcement of energy-efficient building standards.

Better regulation, combined with market-based instruments, may obviate the need for direct investment controls at the center. In the near term investment controls may be needed to maintain macroeconomic stability, but the current investment system inadvertently encourages the building of suboptimal-scale power plants. The investment approval process allows provinces to invest up to $30 million without securing central government approval. Since getting approval often entails delays, power-starved provinces find it easier to install small units than large ones. Phasing out the investment approval system and strictly enforcing emission taxes would discourage the construction of small power plants and create incentives to build more efficient and less polluting plants.

Regional regulation

River basin management is a pressing concern. As things stand, divided jurisdictions across municipalities and provinces have not worked to the environment's advantage. An integrated river basin management system under strong leadership is needed to implement water policies ranging from extraction rights to pollution abatement.

Controlling sulfur emissions is also an important regional issue. The National Air Pollution Prevention and Control Law of 1995 includes abatement plans for regions with high levels of sulfur dioxide and acid rain. Sulfur removal can be costly. It is important that regulation allow plants to choose among abatement measures within the limits of regional control targets.

An alternative to installing sulfur control equipment in all of China's power plants is to concentrate on area sources in regions with high acid rain. A medium-cost investment plan could achieve similar environmental goals—protecting human health, forests, and agriculture—at much lower cost, even while allowing greater emissions than the current target. This strategy would focus on cleaning coal and applying medium-cost technology to new power plants and industrial boilers and, where feasible, using low-sulfur fuels. This targeted, medium-technology strategy promises the largest return to investments.

Urban planning

New regulations at the national and regional levels will only be fully effective if they are accompanied by changes in municipal planning. Urban land use planning is one of China's greatest long-term opportunities. China today is unique in beginning its relationship with the automobile with high urban densities and low levels of car ownership. These two features create the opportunity to develop cities in ways that incorporate efficient and environmentally sustainable transportation. Well-organized cities with good public transit can have half as many automobiles clogging their streets as poorly managed cities with weak public transit. To achieve this goal, transportation planning will have to avail itself of the full range of available tools—from time-of-day tolls to restricted access to downtown areas, from well-conceived parking requirements to better land use planning.

Since the number of automobiles is growing at twice the rate of urban road systems and average driving speeds are slowing, only better planning can channel urban transportation into efficient growth patterns. Planning should be linked to planned relocation of urban industries from downtown areas to industrial parks, away from population concentrations. This move would do much to reduce the exposure of Chinese people to damaging air pollution. Curitiba, Brazil, for

---

BOX 8.2

Markets join national regulators in controlling pollution

A new approach to regulation in Asia is showing that local communities and market forces can be powerful allies in the struggle against industrial pollution. The new programs, called the Program for Pollution Control, Evaluation, and Rating (PROPER) in Indonesia and Ecowatch in the Philippines, rate and publicly disclose factories' environmental performance.

Both programs assign factories a color rating based on the government's evaluation of its environmental performance. A blue rating is given to factories in compliance with regulatory standards, gold is reserved for world-class performers, and black for factories that have made no attempt to control pollution and are causing serious damage. Intermediate ratings are red, for factories that have some pollution control but that fall short of compliance, and green, for factories whose emissions control and housekeeping procedures significantly exceed those needed for compliance. In the future both Indonesia and the Philippines will assign green ratings to factories that meet the environmental conditions of the new ISO 14000 standard (see box 5.3).

The ratings are designed to reward good performance and call public attention to polluters who are not in compliance with regulations. Armed with this information, local communities can negotiate better environmental arrangements with neighboring factories, firms with good performance can advertise their status and earn market rewards, investors can more accurately assess environmental liabilities, and regulators can focus their limited enforcement resources on the worst performers. During its first two years of operation PROPER has proven quite effective in moving poor performers toward compliance. The number of factories receiving a blue rating increased by more than half between June 1995 and September 1996 (see table). PROPER has now moved beyond the pilot stage; 400 factories are currently in the system, and Indonesia plans to extend coverage to 2,000 polluters by the year 2000.

Inspired by these and other examples of public information in action, Colombia and Mexico are now moving rapidly toward their own public disclosure programs.

PROPER considerably boosted factories' compliance with environmental standards
(number of factories)

Rating	June 1995	September 1996
Gold	0	0
Green	5	5
Blue	61	94
Red	115	87
Black	6	1

Source: World Bank staff.

example, began effective urban planning two decades ago, and today gasoline consumption is 30 percent less than in comparable Brazilian cities.

The odds of successful environmental planning and regulation increase with greater community participation in environmental matters. In China community participation is closely correlated with enforcement of pollution levies. In other countries basic information about ambient pollutant concentrations and plants' emissions performance have been instrumental in improving community participation in regulatory enforcement. The government might well consider adopting programs similar to those in Indonesia and the Philippines (box 8.2).

Clear water, blue skies?

China's environment can improve dramatically if assertive policies are adopted promptly and systematically. These policies must encourage conservation and efficiency in resource use while eliciting the investments needed to achieve China's environmental objectives. If they do, China's environment should be as livable as its per capita income suggests (figure 8.1).

China can turn its assets—an increasing market orientation, rapid economic growth, and strong administrative capacity—into advantages for preserving and improving its environment for future generations. To do so, China must harness the market to preserve the environment by ensuring that prices reflect environmental costs. It must harness growth by creating incentives for private investments and by making wise public investments. And it must harness its formidable administrative capacity through better regulations at the

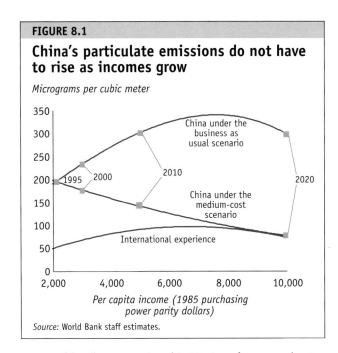

FIGURE 8.1

China's particulate emissions do not have to rise as incomes grow

Micrograms per cubic meter

Source: World Bank staff estimates.

national level, new regional initiatives for water basins and sulfur control, and enhanced urban planning and enforcement at the local level (table 8.2).

These efforts will require some sacrifices in the near term. Investments in the environment will have to be somewhat more than is currently planned, and policy efforts will have to be intensified. But these costs are small relative to the future costs of cleanup if actions are delayed. More important, these sacrifices are small relative to the enormous improvements in the quality of life that will accrue to future generations.

Note

1. Chinese researchers prefer to use the "human capital" approach to mortality and health risk valuation, which produces a damage estimate of about 3.5 percent of GDP (see chapter 2).

TABLE 8.2A
Summary of policy recommendations—Harnessing markets

Short term	Medium term
Energy	
• To encourage more efficient use, remove remaining controls on coal prices and begin phasing out planned coal allocation.	• Complete phaseout of planned coal allocation.
• To encourage shipment of high-quality coal, phase out coal transport subsidies and reduce the administrative allocation of railway capacity.[a]	• Establish a rational pricing framework for coal transport and distribution services.
• To develop alternatives to coal, establish a transparent pricing framework for oil and gas and ease entry to all parts of the oil and gas industry for domestic as well as foreign investors.	
• To spur clean fuel use, reduce price discrimination and increase market allocation of natural gas, and allow prices to reflect consumers' willingness to pay.[a]	• Fully deregulate natural gas production and supply to encourage wider use of natural gas in residential as well as industrial applications.
• To allow for full cost recovery, remove subsidies for coal, gas, and district heating and adjust tariffs.	
• To internalize health costs, introduce a pollution tax based on sulfur and ash content of coal in experimental provinces.	• Generalize sulfur tax based on experiment.
• Apply "polluter pays" principle to car use by increasing gasoline and diesel taxes.	• Complete increase in gasoline and diesel taxes, possibly to levels approximating those in Japan and the Republic of Korea.
Industry and urban transportation	
• To control sulfur in acid rain regions, implement sulfur emission levies for large point sources (such as power plants) and allow adjustment of electricity prices.[a]	• Establish tradable sulfur emission permits for medium-size and large coal users in acid rain regions.
• To increase abatement, redesign pollution levies based on volumes of discharges (rather than concentrations).[a]	
• To induce abatement levels in line with regional pollution control goals, raise pollution levies.	
• To reflect the real cost of automobile transport, apply infrastructure charge to car ownership through sales and licensing taxes.	• Develop tolls, parking fees, and other mechanisms to price car infrastructure use in heavy-traffic urban areas.
Water	
• To reflect the marginal cost of supply, increase water prices, especially for industrial and commercial users, with compensatory measures for low-income households.	• Adjust prices to encourage water conservation.
• To protect water resources from overexploitation, introduce regulations on a raw water extraction tax.[a]	
• To match the costs of self-supply with municipal water supply, raise fees for self-extraction.	
• To cover the full costs of construction and operation, increase sewerage charges and wastewater treatment fees.	
• To improve collections, integrate collection of water and sewerage tariffs.	

TABLE 8.2B

Summary of policy recommendations—Harnessing growth

Short term	Medium term
Energy	
• Remove obstacles to commercial investment, particularly from nonstate firms, in coal washing.	
• Set emission standards for small coal-fired power units at the same levels as for 200 megawatt units and tighten enforcement through factory certification and onsite inspection.	
• Increase natural gas supply by developing domestic reserves and investing in infrastructure for importing piped natural gas and liquefied natural gas.	
• Begin to phase out coal-based fertilizer plants by importing fertilizers, investing in natural gas– or petroleum–based plants, and removing incentives for building and operating coal-based plants.	• Complete phaseout of coal-based fertilizer plants.
• Increase investment in research and development for clean coal technologies and their commercial application.	• Increase investment in research and development for renewable energy sources and their commercial application.
Industry and urban transportation	
• Strengthen programs to promote high-quality and high-efficiency pollution control equipment, especially for particulate control and wastewater treatment.	
• Formulate a public transit–based urban transport investment strategy and create incentives for private participation in the construction and operation of public transit systems.	• Implement public transit–based urban transport plans.
• Establish general framework for financing long-term municipal investments, including water and public transport, through bonds and build-operate-transfer (BOT) schemes. Investments should be limited to creditworthy municipalities with comprehensive public transit strategies.	• Expand municipal governments' access to the domestic bond market based on creditworthiness.
Water	
• Increase investment in water supply and sewerage systems in both rural and urban areas to meet basic drinking and sanitary needs.[a]	
• Provide technical and financial assistance to disseminate water-saving technologies for agriculture and industries.	

TABLE 8.2C

Summary of policy recommendations—Harnessing administrative capacities

Short term	Medium term
Overall	
• Develop long-term environmental master plans to strengthen the management of airsheds, watersheds, and hazardous waste.[a]	• Begin implementing the master plans by phasing in interim targets, emphasizing relocation of polluting industries away from population centers.
• Redesign environmental impact assessments and the Three Synchronizations program to broaden their evaluation basis from discharge standards to local and regional assimilative capacities.[a]	• Gradually detach financial support for local environmental protection bureaus from the levy fund to eliminate the incentive to maintain revenue with less enforcement.
• Promote public awareness of environmental degradation and its consequences through education and open access to information on pollution and environmental quality.[a]	• Fund local environmental protection bureaus from municipal budgets.[a]
• Build a strong knowledge base for more effective environmental management by investing in monitoring and scientific research.[a]	
Energy	
• To reduce environmentally destructive and unsafe mining practices, tighten regulation of small coal mines and collect resource and pollution taxes.	
• Enforce energy efficiency standards for heated residential buildings and formulate energy efficiency standards for other residential buildings and for commercial buildings.[a]	• Enforce energy efficiency standards for all residential and commercial buildings.
• Formulate and implement energy efficiency standards for major household appliances in close collaboration with industry.	
• Implement the lead phaseout plan, first in high-emission cities.[a]	
Industry and urban transportation	
• To remove disincentives to invest in cleaner and energy-efficient production technologies, reform industrial investment approval and lending policies.[a]	• Enact grading system for plants environmental performance.
	• Establish a framework for implementing ISO 14000.
• Incorporate township and village industrial enterprise pollution control into the existing regulatory framework and control highly polluted ones through stricter enforcement and access to abatement fund.	• Designate industrial zones for water-polluting township and village industrial enterprises so that centralized control and abatement technology can be applied.
• Pool pollution levies into a fund not earmarked for specific enterprises, and use the levy fund primarily to support high-return pollution abatement.	• Phase out subsidies for pollution control.
• To contain urban sprawl and avoid traffic congestion, formulate urban land use planning strategies in all medium-size and large cities.	
• Implement progressive emission and fuel efficiency standards for automobiles in close collaboration with the automobile industry.[a]	
Water	
• Strengthen river basin management by consolidating government agencies and allocating responsibilities to fewer institutions.	

a. Initial steps have been taken to adopt and implement this policy.